Table of Contents

Abstract

This report is a translation from the Russian of a survey article by L.D Faddeyev, which appeared in Uspekhi Matem. Nauk., 14, No. 4 (88), p. 57 (1959). Our own interest in this article lies in its relevance to the inverse scattering problem, - that is, the problem of determining information about a medium from which an electromagnetic wave is reflected, given a knowledge of the reflection coefficient. Similar questions concerning scattering phenomena in other branches of physics, e.g., in quantum mechanics, can be investigated by means of the same theory. We have therefore thought it worthwhile to reproduce and distribute the translation. A good indication of the contents is given in the Introduction.

Introduction

This paper is devoted to a survey of the following fundamental problem arising in the quantum theory of scattering: the solution of the equation satisfying

$$L\psi = -\frac{d^2}{dx^2}\psi(x,k) + q(x)\psi(x,k) = k^2\psi(x,k), \qquad (0.1)$$

$$\psi(0,k) = 0 \qquad (0.2)$$

behaves asymptotically like

$$\psi(x,k) \approx C(k)\sin(kx - \eta(k)), \qquad (0.3)$$

provided the potential $q(x)$ decreases sufficiently fast as x tends to infinity; to what extent does assigning the function $\eta(k)$ determine $q(x)$ and how are they related. This problem is one example of the general question concerning the relationship between the S-matrix and the energy operator in scattering theory. The operator L defined by equation (0.1) and condition (0.2) is the simplest example of the energy operator occurring in scattering theory, and the function $S(k) = e^{-2i\eta(k)}$ the simplest example of the S-matrix or scattering operator.

Introduced for the first time by Wheeler, the S-matrix has since been frequently applied in scattering theory following the publication of Heisenberg's papers [1]. In these articles, the following time-independent definition of the S-matrix was given. A wave function, describing the steady state of a system (for simplicity, we restrict ourselves to a system of two particles), has an asymptotic representation in the space variables

$$\Psi\Big|_{r\to\infty} \approx \Psi_1 + \Psi_2 \tag{0.4}$$

where r is the distance between particles. Ψ_1 and Ψ_2 are respectively outgoing and incoming waves. Therefore, Ψ_1 contains the factor e^{ikr} and Ψ_2 is proportional to e^{-ikr}, k being the wave number characterizing the energy of the state Ψ. The quantity relating the amplitudes of these two functions is called the S-matrix. In our illustration,

$$\psi(x,k) \approx \frac{C(k)}{2i}\left[e^{ikx-i\eta(k)} - e^{-ikx+i\eta(k)}\right], \tag{0.5}$$

i.e., the first term corresponds to an outgoing wave, the second to an incoming wave, and their amplitudes are related by the factor

$$S(k) = e^{-2i\eta(k)}. \tag{0.6}$$

This expression constitutes the S-matrix for our example.

Heisenberg's theory of the S-matrix was further developed in the papers of Møller[2] who gave a time-dependent definition of the S-matrix which physically was more meaningful. Since then this time-dependent formulation of the scattering problem has received a great deal of attention (see the appendix) and can be stated in the following way. The energy operator of a system consists of two terms

$$L = L_0 + V, \tag{0.7}$$

where L_0 corresponds to the energy of the free particles and V to the interaction energy. Long before collision, i.e., for negatively infinite time, the state of the non-interacting particles is described by a vector $\Omega_-(t)$* whose dependence on time is determined by the operator L_0:

---

* In conformity with the established convention, a state vector will be some element of Hilbert space in which all operators act.

$$\Omega_-(t) = e^{-iL_o t}\Omega_-, \tag{0.8}$$

Ω_- being a constant vector characterizing the initial state of the system. For finite time, the state $\Omega(t)$ is a solution of the Schroedinger equation

$$i\frac{\partial\Omega(t)}{\partial t} = L\Omega(t) = (L_o + V)\Omega(t), \tag{0.9}$$

and is required to take on the initial state $\Omega_-(t)$ in the sense:

$$\lim_{t\to-\infty} \|\Omega(t) - \Omega_-(t)\| = 0. \tag{0.10}$$

Over a large interval of time after collision, the particle motion again becomes free so that asymptotically as $t \to \infty$

$$\|\Omega(t) - \Omega_+(t)\| \to 0, \tag{0.11}$$

where $\Omega_+(t) = e^{-iL_o t}\Omega_+$. The manner in which the asymptotic state vector changes determines the nature of the scattering process. The operator S which relates the asymptotic vectors Ω_+ and Ω_- according to the formula

$$\Omega_+ = S\Omega_-, \tag{0.12}$$

is called the <u>scattering operator</u> or S-matrix.

In Sec. 3, it will be shown that this formulation holds for the example in question, the S-matrix being given by the function $S(k) = e^{-2i\eta(k)}$ occurring in the time-independent definition. This fact typifies a general aspect of the stationary and nonstationary formulations of the S-matrix in case both definitions are valid.

Heisenberg came to consider the S-matrix as a means of overcoming the difficulties encountered in modern relativistic theory of elementary particles. He felt it was necessary to introduce a new fundamental constant having dimension length. Therefore, he analyzed the current theory and rejected as unobservable those notions which contradicted the idea of a fundamental length. Only those quantities experimentally observable would be put in a future theory. In this sense, the S-matrix satisfies the requirements of Heisenberg. It depicts the wave function at large distances and is thus not contrary to the fundamental length hypothesis. Moreover, the scattering cross-sections, which can be directly measured experimentally, are expressed in terms of the elements of the S-matrix. Heisenberg also conjectured that the discrete energy levels corresponding to bound states of the particles should be determined by the analytic continuation of the S-matrix into the complex energy plane.

Connected with Heisenberg's supposition that the S-matrix was more fundamental than the Hamiltonian was the question of clarifying the relationship between these two characterizations of a system. In particular, in what sense should one define the Hamiltonian on the basis of the S-matrix when both of these notions were used in a theory. In addition to the theoretical aspect, the inverse problem, i.e. the reconstruction of the energy operator from its S-matrix, could be of great practical value in the interpretation of experimental scattering data as well as in the determination of various properties of the particles which are not directly measurable.

The simplest example in scattering theory is the radial equation which describes the scattering by a fixed spherically symmetric center:

$$-\frac{d^2}{dx^2}\psi(x,k)+\left(\frac{\ell(\ell+1)}{x^2}+q(x)\right)\psi(x,k)=k^2\psi(x,k). \qquad (0.13)$$

This reduces to the case already mentioned for $\ell = 0$. The first attempts at solving the inverse problem were undertaken by Frødberg[3] and Hylleraas[4]. They worked out a formal procedure using series whose convergence was very likely. However, Bargmann[5] constructed explicit examples in which different potentials gave rise to the same S(k) and to the same discrete energy levels. This showed that a potential which led to given energy levels and scattering functions S(k) could not be reconstructed in a unique manner. Levinson[6] showed that this lack of uniqueness must be related to the existence of a discrete spectrum. To wit, he proved that the reconstructed potential is unique when there is no discrete spectrum. The precise mathematical reason for this was explained by Marchenko[7] who showed that the S-function* determines the continuous portion of the so-called spectral function of equation (0.1). To find the spectral function when there is a discrete spectrum, it is necessary to give not only the location of the eigenvalues but also, for example, the derivatives of the corresponding normalized eigenfunctions at x = 0. Marchenko[8] also showed that the spectral function determines a potential uniquely. Thus, Marchenko related the problem in question to the so-called inverse Sturm-Liouville problem which already had been discussed in the mathematical literature. An analogous result was obtained at approximately the same time by Borg[66]. By developing Levinson's method, Jost and Kohn[9], [10] independently came to the same conclusion concerning the reason for this lack of uniqueness. In addition to the original potential q(x), they gave an explicit formula for a family of potentials having the same

---

* We will call the scattering operator in our example the S-function.

S-function and the same discrete energy levels. An analogous formula was obtained by Holmberg[11].

A procedure for explicitly constructing a potential without the singularity $\ell(\ell + 1)/x^2$ from its spectral function was formulated by Gelfand and Levitan[12]. They reduced the problem to a linear integral equation and gave sufficient conditions in terms of the spectral function quaranteeing that it be the spectral function of some equation with a potential from a given class. The results of Gelfand and Levitan on the inverse Sturm-Liouville problem were immediately applied to the inverse scattering problem by Jost and Kohn[13] and Levinson[14]. In [13], a formula was given for a family of equivalent potentials having one and the same S-function and discrete energy levels. Stricter conditions (both necessary and sufficient) on the spectral function were obtained by Krein[15]. His paper completed the general problem of reconstructing equation (0.1) from its spectral function. However, since the passage from the S-function to the spectral function was not entirely trivial, there still remained the unanswered question concerning the nature of the set of possible S-functions corresponding to the potentials from a given class. This problem was solved by Krein[16] and Marchenko[17] who gave a series of conditions in terms of the Fourier transform of the function $S(k) - 1$. Marchenko showed that the potential $q(x)$ possesses the same properties for x tending to zero and infinity as does the derivative of this Fourier transform. The final inequalities deduced by Marchenko permitted him to formulate necessary and sufficient conditions on the S-function ensuring that a potential from a given class would correspond to it.

After the basic papers of Gelfand and Levitan, Krein and Marchenko, a great deal of work was devoted to carrying over their results to an equation containing the singular term $\ell(\ell + 1)/x^2$, an equation over the interval $-\infty < x < \infty$, a system of equations, and the relativistic equations. A brief survey is given in the appendix.

It is interesting to note that in the USSR the inverse problem has been studied on the whole by mathematicians whereas abroad almost exclusively by physicists who merely use the method of Gelfand-Levitan as interpreted by Levinson, Jost and Kohn. An explanation of the general features of this method, and its application to the solution of various problems, was undertaken in a series of papers by Kay and Moses [18] - [21]. These authors used the general concept of transformation operators developed by Friedrichs [22], [23].

Recently work has been devoted to applying the results of the inverse problem in the interpretation of experimental scattering data [24], [25], [26].

Thus, the inverse scattering problem for the simplest example of the radial equation was solved during the course of about a decade, and a large amount of literature has been devoted to it. In the present survey, we will endeavor to give the results of a majority of these papers in their most general form. This will make clearer the way in which the basic results are carried over to other problems. In this, the general approach to transformation operators developed by Friedrichs and applied to the inverse problem by Kay and Moses will play an essential role.

All of the basic results in the inverse scattering problem could be obtained using one of the methods of Gelfand-Levitan, Marchenko or Krein. Our presentation will not stick to any particular one of these methods, but rather at different points will make use of different methods. We will attempt to establish their connection considering that each one of these methods explains different aspects of the mathematical structure of the entire problem.

Due to the large material content, all mathematical proofs will not be carried out in a completely rigorous fashion. Many of our considerations will be of a heuristic nature where the justification of details would require greater background than in other more standard proofs. We will nevertheless use these heuristic proofs in order not to obscure the conceptual side of this work with lengthy mathematical arguments. We are confident that the physicist-reader will find our discussions completely convincing and that the mathematician will be able to reconstruct the deficient proofs so as to make them completely rigorous. On the other hand, we have tried to state theorems in their most precise form.

Let us briefly outline the basic ideas and plan of the survey. The first 13 sections are devoted to the solution of the inverse scattering problem for the operator L defined by equation (0.13) and the condition $\psi(0) = 0$ without the singularity $\ell(\ell + 1)/x^2$. We treat the operator L as a perturbation of the operator L_0 defined by the differential expression $L_0\psi = - d^2\psi(x)/dx^2$ and the same condition $\psi(0) = 0$. According to Friedrichs, a transformation operator U is defined as the solution of the operator equation

$$LU = UL_0. \qquad (0.14)$$

This implies that any transformation operator which has an inverse generates a similarity transformation of the perturbed operator into the unperturbed:

$$U^{-1}LU = L_o. \qquad (0.15)$$

This transformation operator U replaces the eigenfunctions of the continuous spectrum of the operator L in all considerations. Roughly speaking, its kernel is obtained by expanding the eigenfunctions of the continuous spectrum of L in terms of the eigenfunctions of the operator L_o.

In Sec. 4 and Sec. 5, it is shown that such transformation operators exist for our example and that the completeness theorem for the eigenfunctions of L, proven in Sec. 2, can be written down in terms of the transformation operator in the form

$$UWU^* = I, \qquad (0.16)$$

(for simplicity, we have restricted ourselves here to the case when L has a discrete spectrum; in the text, this restriction is not imposed). Here, W is a positive-definite self-adjoint operator commuting with L_o. W determines the 'normalization' of the corresponding operator U.

A characteristic feature of our example is that among the transformation operators there exist Volterra operators of the form

$$U_B f(x) = (I + K)f(x) = f(x) + \int_0^x K(x,y)f(y)dy. \qquad (0.17)$$

The operator W corresponding to U_B is constructed using

$$W(k) = \frac{1}{M(k)M(-k)}, \qquad (0.18)$$

where M(k) is a certain function introduced in Sec. 1. One might call $M(\sqrt{\lambda})$ the determinant of the operator L - λL. In fact, in Sec. 2 it is shown that this function appears in the denominator of the resolvent kernel

The subscript B used throughout and which transliterates into English as V stands, of course, for Volterra.

of L and determines its singularities. These consist of a branch cut and poles corresponding respectively to the continuous spectrum and points of the discrete spectrum.

In Sec. 3, it is shown that the time-dependent formulation of the scattering problem is valid for our example provided that L_0 is taken as the energy operator of the free particles. The corresponding scattering operator is determined by using the function

$$S(k) = \frac{M(-k)}{M(k)} . \tag{0.19}$$

In Sec. 6, it is shown how to establish the relationship between W(k) and S(k) with the help of (0.18) and (0.19).

In Sec. 8, on the basis of (0.16) and the triangularity of the kernel K(x,y), a linear integral equation is obtained connecting the kernels of W and K. In Sec. 9, this equation is studied and the inverse problem is solved for the case when L doesn't have a discrete spectrum. Supplementary facts necessary for a treatment of the general case are cited in Sec. 12.

The approach described corresponds to the Gelfand-Levitan method. Another procedure, related to Marchenko's method, is based on the application of the operator $V_B = I + A$ introduced in Secs. 4 and 7:

$$V_B f(x) = f(x) + \int_x^{\infty} A(x,y)f(y)dy. \tag{0.20}$$

This operator is not a transformation operator in the sense of the general definition. However, its relationship to the transformation operator $\tilde{U}_B = U_B W = (U_B^*)^{-1}$ is established in Sec. 7. From this relation and (0.16), it follows that the next identity holds for the operator V_B:

$$V_B(I - F)V_B^* = I, \tag{0.21}$$

where the operator F can be constructed directly using the function S(k). By means of (0.21), a linear integral equation is deduced which relates the kernel A(x,y) to the function S(k), thus permitting one to solve the inverse problem. This integral equation is used in Sec. 10 to investigate the connection between S(k) and q(x). Several aspects of Krein's method are illustrated in Sec. 11. In Sec. 13, the construction of some operator L from a known operator L_1 when the S-function of L differs from that of L_1 by a rational factor is considered. This is an important problem in applications. In Sec. 14 and 15, the results deduced are carried over to the radial equation (0.13) when $\ell > 0$.

In order not to interrupt the presentation, we will not refer to original papers in the text. The literature is cited in a special appendix. A series of comments are made there and a brief survey is given of work done on the inverse scattering problem which is not included in the text.

For her valuable hints and corrections, the author expresses his deepest thanks to O. A. Ladizhenskaya who read through the survey in manuscript. He would also like to thank Z. S. Agranovich and V. A. Marchenko for their courtesy in allowing him to look at their monograph prior to its publication.

1. The solutions $\varphi(x,s)$, $f(x,s)$ and their relationship, existence and inequalities The function $M(s)$ and its properties.

In this section, some basic features of solutions of the equation

$$-y'' + q(x)y = s^2 y, \quad s = \sigma + i\tau, \tag{1.1}$$

are assembled which will be utilized in the subsequent presentation. In all lemmas, it is assumed without further mention that $q(x)$ is a locally summable function and satisfies the condition

$$\int_0^\infty x|q(x)|dx = C < \infty . \tag{1.2}$$

The solutions $\varphi(x,s)$ and $f(x,s)$ are determined by the conditions:

$$\varphi(x,s): \quad \varphi(0,s) = 0, \quad \varphi'(0,s) = 1, \tag{1.3}$$

$$f(x,s): \quad \lim_{x \to \infty} e^{-isx} f(x,s) = 1. \tag{1.4}$$

Equation (1.1) and the conditions (1.3) and (1.4) are equivalent to the following integral equations:

$$\varphi(x,s) = \frac{\sin sx}{s} + \int_0^x \frac{\sin s(x-t)}{s} q(t)\varphi(t,s)dt, \tag{1.5}$$

$$f(x,s) = e^{isx} + \int_x^\infty \frac{\sin s(t-x)}{s} q(t)f(t,s)dt, \tag{1.6}$$

which can be obtained by the method of variation of parameters. With the help of these equations the following lemmas are proven:

Lemma 1.1 *For each $x \geq 0$, $\varphi(x,s)$ is an entire function of s for which the estimate**

$$|\varphi(x,s)| \leq K \frac{xe^{|\tau|x}}{1+|s|x} \tag{1.7}$$

holds. Moreover, $\varphi(x,s)$ is an even function of s for real s.

Lemma 1.2 *For each $x \geq 0$, $f(x,s)$ is analytic with respect to s in the upper halfplane $\tau > 0$ and continuous down to the real axis. Moreover, the inequality*

$$|f(x,s)| \leq Ke^{-\tau x}, \qquad \tau \geq 0 \tag{1.8}$$

holds.

Lemma 1.3 $f(x,s)$ *satisfies the following inequalities:*

$$|f(x,s) - e^{isx}| \leq K \frac{e^{-\tau x}}{|s|} \int_x^\infty |q(t)|dt, \qquad \tau \geq 0, \tag{1.9}$$

$$|f(x,s) - e^{isx}| \leq Ke^{-\tau x} \int_x^\infty t|q(t)|dt, \qquad \tau \geq 0, \tag{1.10}$$

$$|f'(x,s) - ise^{isx}| \leq Ke^{-\tau x} \int_x^\infty |q(t)|dt, \qquad \tau \geq 0. \tag{1.11}$$

The estimate (1.9) is convenient for $|s| \to \infty$ and can be applied when $x \neq 0$. The estimates (1.10) and (1.11) are convenient for $x \to \infty$. In addition, from (1.11) one can conclude that

$$\lim_{x \to 0} xf'(x,s) = 0. \tag{1.12}$$

---

* All absolute constants depending only on C (which can be different) will be denoted by K.

In fact, as $x \to 0$,

$$x \int_x^\infty |q(t)|dt \leq \int_x^{\sqrt{x}} t|q(t)|dt + \sqrt{x} \int_{\sqrt{x}}^\infty t|q(t)|dt \to 0.$$

Lemma 1.4 <u>The function $f(x,s)$ for any x is continuously differentiable with respect to s down to the line $\tau = 0$ with perhaps the exception of the point $s = 0$. The estimate</u>

$$|\dot{f}(x,s) - ixe^{isx}| \leq \frac{K}{|s|} e^{-\tau x}, \quad \tau \geq 0, \tag{1.13}$$

<u>holds uniformly in x.</u>

Lemma 1.5 <u>For large</u> $|s|$

$$\varphi(x,s) = \frac{\sin sx}{s} + o\left(\frac{e^{|\tau|x}}{|s|}\right), \tag{1.14}$$

$$f(x,s) = e^{isx} + o(e^{-\tau x}), \quad \tau \geq 0, \tag{1.15}$$

<u>uniformly for all $0 \leq x$.</u>

When s is real, it is not difficult to establish the relationship between $\varphi(x,s)$ and $f(x,s)$. Without further mention, we will write k instead of s when s is real. The solutions $f(x,k)$ and $f(x,-k) = \overline{f(x,k)}$ for $k \neq 0$ are linearly independent solutions of equation (1.1). In fact, their Wronskian

$$[f(x,k);f(x,-k)] = f'(x,k)f(x,-k)-f(x,k)f'(x,-k) = 2ik \tag{1.16}$$

is not zero.

In consequence of the realness of $\varphi(x,k)$

$$\varphi(x,k) = \frac{1}{2ik}\left(f(x,k)\overline{M(k)} - f(x,-k)M(k)\right), \tag{1.17}$$

where $M(k)$ can be determined with the help of the Wronskian (see (1.12)):

$$M(k) = \left[\varphi(x,k);\ f(x,k)\right] = \lim_{x \to 0} \left[\varphi(x,k);\ f(x,k)\right] = f(0,k) \tag{1.18}$$

From this and lemma 1.2, we conclude that $M(k)$ is the limiting value of the function $M(s) = f(0,s)$, analytic in the upper halfplane and such that $M(k) = \overline{M(-k)}$. Let us introduce the notations·

$$A(k) = |M(k)|, \quad \eta(k) = \arg M(k), \tag{1.19}$$

$$A(k) = A(-k), \quad \eta(k) = -\eta(-k). \tag{1.20}$$

Then from lemma 1.4, we infer that for large x

$$\varphi(x,k) = \frac{A(k)}{k}\sin(kx - \eta(k)) + o(1), \tag{1.21}$$

$$\varphi'(x,k) = A(k)\cos(kx - \eta(k)) + o(1). \tag{1.22}$$

It is therefore natural to call $A(k)$ the <u>limiting amplitude</u> and $\eta(k)$ the <u>limiting phase</u>.

Let $\tau > 0$. From equation (1.1) for $f(x,s)$ and the equation

$$-\dot{f}''(x,s) + q(x)\dot{f}(x,s) = 2sf(x,s) + s^2\dot{f}(x,s) \tag{1.23}$$

for $\dot{f}(x,s) = df(x,s)/ds$, it is not difficult to obtain the following identities:

$$f'(0,s)\overline{f(0,s)} - f(0,s)\overline{f'(0,s)} = 4i\sigma\tau \int_0^\infty |f(t,s)|^2 dt, \tag{1.24}$$

$$\dot{f}'(0,s)f(0,s) - f(0,s)f'(0,s) = 2s \int_0^\infty f^2(t,s)dt. \tag{1.25}$$

From the first one, we conclude that $M(s)$ can vanish only for $\sigma = 0$ or $\tau = 0$. The second possibility, however, is excluded by the fact that if $M(k) = 0$ on the real axis, (1.17) would imply that $\varphi(x,k) \equiv 0$, and this is impossible.

In the following, it will be assumed that $M(0) \neq 0$. The vanishing of $M(s)$ for $s = 0$ is equivalent to the solution of $-y'' + q(x)y = 0$, $y(0) = 0$ being bounded as $x \to \infty$ and this happens only in exceptional situations. A treatment of the case $M(0) = 0$ does not present any essential difficulties but only complicates the formulation and proof of theorems.

There still remains the possibility that $M(s) = 0$ for $\sigma = 0$ and $\tau > 0$. From the estimate

$$M(s) = 1 + o(1) \tag{1.26}$$

for large $|s|$ which follows from formula (1.14), we conclude that $M(s)$ can only have a finite number of zeros of the form $s_n = i\kappa_n$ $(n = 1,\ldots,m)$ on the imaginary axis. When $s = s_n$, the solutions $\varphi(x,s_n)$ and $f(x,s_n)$ satisfy the same boundary condition at $x = 0$ and therefore they are proportional

$$f(x,s_n) = f'(0,s_n)\varphi(x,s_n). \tag{1.27}$$

From this and (1.25), it follows that

$$\int_0^\infty \left[\varphi(x,s_n)\right]^2 dx = -\frac{\dot{M}(s_n)}{2s_n f'(0,s_n)}, \tag{1.28}$$

and this implies, in particular, that all the zeros of $M(s)$ are simple. The above results can be summarized in the following form.

Lemma 1.6 <u>The function $M(s)$ is analytic in the upper halfplane and has there a finite number of simple zeros $s_n = i\kappa_n$, $\kappa_n > 0$, $(n = 1,\dots,m)$. For large $|s|$, the estimate (1.26) holds. The function $\dot{M}(s)$ is continuous down to the real axis with the possible exception of the point $s = 0$. Furthermore, $sM(s)$ is continuous everywhere.</u> This last assertion follows from lemma 1.4.

2. Expansion Theorem

The differential equation (1.1) together with the boundary condition defines a self-adjoint operator in $\mathcal{L}_2(0,\infty)$. This operator can be gotten by the extension of the symmetric operator, defined by (1.1), acting on the twice-continuously differentiable functions satisfying the boundary condition and vanishing identically outside some finite interval. We will denote this operator by L.

Consider the kernel

$$\left.\begin{aligned} R_\lambda(x,y) &= \varphi(x, \sqrt{\lambda},)\,\frac{f(y, \sqrt{\lambda})}{M(\sqrt{\lambda})}, \quad x < y, \\ R_\lambda(x,y) &= R_\lambda(y,x), \\ 0 &\leq \arg \sqrt{\lambda} \leq \pi, \end{aligned}\right\} \tag{2.1}$$

which is defined for all complex λ with the exception of a finite number of points on the negative real axis corresponding to the zeros of $M(\sqrt{\lambda})$. By virtue of (1.18), it is not difficult to verify that the kernel $R_\lambda(x,y)$

is a solution of the equation

$$\left(-\frac{d^2}{dx^2} + q(x)\right) R_\lambda(x,y) - \lambda R_\lambda(x,y) = \delta(x - y), \tag{2.2}$$

and also satisfies the boundary conditions:

$$R_\lambda(0,y) = R_\lambda(x,0) = 0. \tag{2.3}$$

In consequence of (1.7) and (1.8), we have

$$|R_\lambda(x,y)| \leq K \frac{x}{1 + |\sqrt{\lambda}|x} e^{-\tau|x-y|}, \tau = \operatorname{Im} \sqrt{\lambda} > 0 \tag{2.4}$$

for complex λ and, hence, the kernel $R_\lambda(x,y)$ determines in $\mathcal{L}_2(0,\infty)$ a bounded operator, namely, the resolvent operator

$$R_\lambda = (L - \lambda I)^{-1}. \tag{2.5}$$

The singularities of the resolvent R_λ in the complex λ-plane consist of a cut along the positive real axis and a finite number of simple poles $\lambda_n = -\kappa_n^2$ $(n = 1,\ldots,m)$ on the negative real axis. The continuous and discrete portions of the spectrum correspond to the cut and poles, respectively. The jump in the resolvent across the cut and the residues at the poles determine the spectral function of the operator L. We come now to the following completeness theorem for the eigenfunctions of the operator L.

Theorem 2.1. The functions $\varphi(s,k)$ $(k \geq 0)$ and $\varphi_n(x) = \varphi(x,i\kappa_n)$ form a complete orthogonal set of functions. The completeness relationship is given by

$$\sum_{n=1}^{m} C_n\varphi_n(x)\varphi_n(y) + \frac{2}{\pi}\int_0^\infty \varphi(x,k)\,\frac{1}{M(k)M(-k)}\,\varphi(y,k)k^2dk = \delta(x-y) \tag{2.6}$$

in which $C_n = 2i\kappa_n f'(0,i\kappa_n)/\dot{M}(i\kappa_n)$ (see (1.28)).

Formula (2.6) can be deduced without any reference to the general aspects of operator theory. Let $f(x)$ be a twice continuously differentiable function vanishing for large x and in the neighborhood of $x = 0$. Then

$$g(x) = -f''(x) + q(x)f(x) \tag{2.7}$$

is continuous and vanishes identically outside some finite interval not containing the origin. From (2.2) and (2.7), it follows that

$$\int_0^\infty R_\lambda(x,y)f(y)dy = -\frac{1}{\lambda}f(x) + \frac{1}{\lambda}\int_0^\infty R_\lambda(x,y)g(y)dy. \tag{2.8}$$

If we take the integral of both sides of (2.8) along a large circle $|\lambda| = N$, the contribution from the second term on the right side of (2.8) as $N \to \infty$ will be zero by virtue of (1.14), (1.15) and (1.26). Thus we get

$$\lim_{N\to\infty} \frac{1}{2\pi i}\oint_{\lambda=|N|}\left[\int_0^\infty R_\lambda(x,y)f(y)dy\right]d\lambda = -f(x). \tag{2.9}$$

On the other hand, if we take the integral of the left side of (2.8) along a path γ consisting of a curve encompassing the cut on the real axis and a large circle $|\lambda| = N$, we obtain

$$\frac{1}{2\pi i}\oint_{\gamma}\left\{\int_0^\infty R_\lambda(x,y)f(y)dy\right\}d\lambda = \frac{1}{2\pi i}\oint_{|\lambda|=N}\left\{\int_0^\infty R_\lambda(x,y)f(y)dy\right\}d\lambda +$$

$$+\frac{1}{2\pi i}\int_0^\infty d\sigma\left\{\int_0^\infty\left[R_{\sigma+i0}(x,y) - R_{\sigma-i0}(x,y)\right]f(y)dy\right\} =$$

$$= \sum_{n=1}^{m} \text{Res}\left\{\int_0^\infty R_\lambda(x,y)f(y)dy\right\}\Bigg|_{\lambda=\lambda_n}.$$

By taking into account (2.1) and (1.17) and passing to the limit as $N\to\infty$, we find on the basis of (2.9)

$$f(x) = \frac{2}{\pi}\int_0^\infty k^2dk\left\{\int_0^\infty \varphi(x,k)\,\frac{1}{M(k)M(-k)}\,\varphi(y,k)f(y)dy\right\} +$$

$$+\sum_{n=1}^{m}\int_0^\infty C_n\varphi_n(x)\varphi_n(y)f(y)dy$$

Finally, by virtue of the fact that the functions $f(x)$ are dense in $\mathcal{L}_2(0,\infty)$, formula (2.6) is obtained.

The functions $\psi^{(+)}(x,k) = \varphi(x,k)/M(k)$ and $\psi_n(x) = \sqrt{C_n}\,\varphi_n(x)$ form an orthonormal system. However, the functions $\psi^{(+)}(x,k)$ are not square-integrable and, hence, are not elements of Hilbert space and eigenfunctions in the usual sense. To attach meaning to them while still remaining in the framework of Hilbert space, one may treat them as kernels of transformations, which diagonalize the operator L. Thus, the transformation

$$T^{(+)}g = G \cdot \quad G(k) = \int_0^\infty g(x)\psi^{(+)}(x,k)dx \tag{2.10}$$

carries any function $g(x)$ in $\mathcal{L}_2(0,\infty)$ into a function $G(k)$ for which $\int_0^\infty |G(k)|^2 k^2 dk < \infty$. Moreover, if $Lg(x)$ belongs to $\mathcal{L}_2(0,\infty)$, then $Lg(x)$ goes into $k^2G(k)$ and the integral $\int_0^\infty |G(k)k^2|^2 k^2 dk$ converges. In the following, the space of functions $G(k)$ with the scalar product

$$(G,G_1) = \frac{2}{\pi}\int_0^\infty \overline{G(k)}G_1(k)k^2dk \tag{2.11}$$

will be denoted by $\mathcal{L}_k$, and the space of square-integrable functions $g(x)$, which was previously called $\mathcal{L}_2(0,\infty)$, will be denoted by $\mathcal{L}_x$. The transformation $T^{(+)}$ acts from $\mathcal{L}_x$ into $\mathcal{L}_k$. The adjoint transformation $T^{(+)*}$ acts from $\mathcal{L}_k$ into $\mathcal{L}_x$ according to the formula

$$g = T^{(+)*}G: \quad g(x) = \frac{2}{\pi}\int_0^\infty G(k)\overline{\psi^{(+)}(x,k)}k^2dk. \tag{2.12}$$

The orthogonality of the eigenfunctions $\psi^{(+)}(x,k)$ can thus be expressed by using the new terminology in the following way:

$$T^{(+)}T^{(+)*} = I^k. \tag{2.13}$$

Here, I^k denotes the identity operator in $\mathcal{X}_k$. Relation (2.6) can now be written down in the following form:

$$T^{(+)*}T^{(+)} = I_c^x. \tag{2.14}$$

Here, I_c^x is the projection operator onto the proper subspace of the operator L corresponding to the continuous spectrum. The superscript x indicates that this operator acts on $\mathcal{X}_x$. A transformation of the type $T^{(+)}$ can be associated with any solution of equation (1.1) satisfying a zero boundary condition. Let $X(x,k)$ be solution of (1.1) such that $X(0,k) = 0$ and $X'(0,k) \neq 0$ for all k. Consider the transformation

$$Tg = G: \quad G(k) = \int_0^\infty g(x)X(x,k)dx. \tag{2.15}$$

The transformations T and $T^{(+)}$ are related by the formula:

$$T = N^k T^{(+)} \tag{2.16}$$

where N^k is a 'normalizing factor'; N^k is an operator in $\mathcal{X}_k$, which multiplies by the function $N(k) = X'(0,k)/\psi^{(+)'}(0,k)$. The completeness relation and orthogonality condition can be written down in terms of T in the following fashion:

$$T^*W^kT = I_c^x, \quad TT^*W^k = I^k, \tag{2.17}$$

where

$$W^k = (N^k)^{-1}(N^k*)^{-1}. \tag{2.18}$$

The somewhat formal arguments at the end of this section will turn out to be useful in Sec. 5.

3. Asymptotic behavior of the solution of the Shroedinger equation for large time.

The expansion theorem for the operator L deduced in the preceding section enables one to solve by Fourier methods time-dependent equations involving the operator L. We will, in particular, be interested in the behavior of solutions of the Schroedinger equation

$$i\,\frac{\partial f(x,t)}{\partial t} = Lf(x,t), \quad f(x,t)\Big|_{t=o} = f_o(x) \tag{3.1}$$

for large $|t|$.

By expanding $f_o(x)$ with respect to the eigenfunctions of L, one can represent the solution of (3.1) in the following way:

$$f(x,t) = \frac{2}{\pi}\int_0^\infty F(k)\,\overline{\psi(x,k)}\,e^{-ik^2t}k^2dk + \sum_{n=1}^{m} F_n\,\overline{\psi_n(x)}\,e^{i\kappa_n^2t}. \tag{3.2}$$

Here $\psi(x,k)$, $\psi_n(x)$ form an arbitrary set of orthonormal eigenfunctions of L $\left[\psi(x,k)\right.$ and $\psi_n(x)$ may differ from the functions $\psi^{(+)}(x,k)$ and $\psi_n(x)$ discussed in Sec. 2 by a factor of modulus one$\left.\right]$ and

$$F(k) = \int_0^\infty f_o(x)\psi(x,k)dx, \quad F_n = \int_0^\infty f_o(x)\psi_n(x)dx. \tag{3.3}$$

In particular, one may take as the set $\psi(x,k)$ the functions $\psi^{(+)}(x,k)$ or $\psi^{(-)}(x,k) = \psi^{(+)}(x,k)M(k)/M(-k) = \overline{\psi^{(+)}(x,k)}$. The behavior of $f(x,t)$ for large $|t|$ can be analyzed on the basis of the following lemma:

Lemma 3.1. <u>Let</u> $F(k)$ <u>be an arbitrary function in</u> $\mathcal{L}_k$, <u>i.e.</u>

$$\int_0^\infty |F(k)|^2 k^2 dk < \infty , \tag{3.4}$$

<u>and set</u>

$$f^{(\pm)}(x,t) = \int_0^\infty F(k)\, \overline{\psi^{(\pm)}(x,k)}\, e^{-ik^2t}\, k^2 dk, \tag{3.5}$$

$$g(x,t) = \int_0^\infty F(k)\, \frac{\sin kx}{k}\, e^{-ik^2t}\, k^2 dk. \tag{3.6}$$

<u>Then</u>

$$\lim_{t\to\pm\infty} \int_0^\infty |f^{(\pm)}(x,t) - g(x,t)|^2 dx = 0. \tag{3.7}$$

It is sufficient to prove this assertion for a set of functions $F(k)$ which is dense in $\mathcal{L}_k$, for an arbitrary function, the theorem then follows by completion. We will assume that $F(k)$ is differentiable and does not vanish in the interval $0 < \alpha \leqq k \leqq \beta < \infty$. The intervals of integration in (3.5) and (3.6) are then finite and do not contain the point $k = 0$. For definiteness, we will assume that $t \to \infty$. By virtue of the fact that $\psi^{(+)}(x,k)$ and $(\sin kx)/k$ are uniformly bounded with respect to all x in $0 < x < \infty$ and all k in $[\alpha,\beta]$, the functions $f^{(+)}(x,t)$ and $g(x,t)$ tend to zero as $t \to \infty$ uniformly in x. Thus the integral in (3.7) vanishes as $t \to \infty$ for any finite interval of integration. We still have to show that $\int_A^\infty |f^{(+)} - g|^2 dx$ converges for arbitrary A.

From (1.17) and the definition of $\psi^{(+)}(x,k)$,

$$\overline{\psi^{(+)}(x,k)} - \frac{\sin kx}{k} = -\frac{1}{2ik} e^{-ikx}\left[\frac{M(k)}{M(-k)} - 1\right] + R(x,k), \tag{3.8}$$

where for $x > 0$

$$|R(x,k)| \leq K \frac{1}{k^2} \int_x^\infty |q(t)|dt \tag{3.9}$$

by virtue of (1.9) and (1.8). In consequence of this estimate,

$$\int_A^\infty \left| \int_\alpha^\infty F(k)R(x,k)e^{-ik^2t}k^2dk \right|^2 dx \leq$$

$$\leq K \int_A^\infty dx \left(\int_x^\infty |q(t)|dt \right)^2 \int_\alpha^\infty |F(k)|^2 k^2 dk \int_\alpha^\infty \frac{dk}{k^2} \leq$$

$$\leq K' \int_A^\infty |q(t)|dt \int_A^\infty t|q(t)|dt \tag{3.10}$$

and for sufficiently large A the integral containing $R(x,k)$ can be made as small as desired uniformly in t. We still must look at the behavior of the integral of the basic term of (3.8)

$$Q_A(t) = \int_A^\infty dx \left| \int_\alpha^\infty G(k)e^{-ikx}e^{-ik^2t}dk \right|^2, \tag{3.11}$$

as $t \to \infty$. The function $G(k) = \frac{F(k)}{2ik}\left[\frac{M(k)}{M(-k)} - 1\right]k^2$ by virtue of lemma 1.6 is finite and continuously differentiable. Now $Q_A(t) = \lim_{B \to \infty} Q_A^B(t)$ where

$$Q_A^B(t) = \int_A^B dx \left| \int_\alpha^\beta G(k) e^{-ikx} e^{-ik^2 t} dk \right|^2$$

$$= \int_\alpha^\beta dk \int_\alpha^\beta d\ell \left\{ G(k)\overline{G(\ell)} e^{-i(k^2-\ell^2)t} \frac{e^{-i(k-\ell)A} - e^{-i(k-\ell)B}}{i(k-\ell)} \right\} = J_A(t) - J_B(t).$$

Here

$$J_B(t) = \int_\alpha^\beta dk \int_\alpha^\beta d\ell G(k)\overline{G(\ell)} \frac{e^{-i(k^2-\ell^2)t} e^{-i(k-\ell)B}}{i(k-\ell)},$$

and because of the singularity in the denominator, the inner integral is taken as a principal value, well-defined in consequence of the differentiability of $G(k)$.

We transform $J_B(k)$ to the following form:

$$J_B(t) = \int_\alpha^\beta dk G(k) \left[\int_\alpha^\beta d\ell \frac{\overline{G(\ell)} e^{-i(k^2-\ell^2)t} - \overline{G(k)}}{i(k-\ell)} e^{-i(k-\ell)B} + \right.$$

$$\left. + \overline{G(k)} \int_\alpha^\beta \frac{e^{-i(k-\ell)B}}{i(k-\ell)} d\ell \right]. \qquad (3.12)$$

The first term in the integrand is continuous for $k = \ell$ and vanishes as $B \to \infty$ by the Riemann-Lebesgue theorem. The integral of the second term tends to π as $B \to \infty$ and therefore, we find that $\lim_{B \to \infty} J_B(t)$ does not depend on t and is equal to $\pi \int_\alpha^\beta |G(k)|^2 dk$.

$J_A(t)$ can also be represented as the sum of two terms one of which vanishes as $t \to \infty$ uniformly in A. The other term has a finite limit independent of A as $t \to \infty$ equal to $\pi \int_{\alpha}^{\beta} |G(k)|^2 dk$. From this it follows that $Q_A(t) \to 0$ and the lemma is therefore proven.

Let us now return to the investigation of the behavior of the solution $f(x,t)$ of the Schroedinger equation. If $f_0(x)$ is orthogonal to the eigenfunctions of the discrete spectrum of L, the sum in the second term of (3.2) does not appear. Let

$$F^{(\pm)}(k) = \int_0^\infty f_0(x)\psi^{(\pm)}(x,k)dx. \tag{3.13}$$

These are the functions which occur in formula (3.2) for $f(x,t)$, the $\psi^{(\pm)}(x,k)$ having been selected for the functions $\psi(x,k)$. It is evident that

$$g^{(\pm)}(x,t) = \frac{2}{\pi} \int_0^\infty F^{(\pm)}(k) \frac{\sin kx}{k} e^{-ik^2t} k^2 dk \tag{3.14}$$

is a solution of the Schroedinger equation with an operator L_0, associated with the equation $L_0 y = -y'' = k^2 y$ and the boundary condition $y(0) = 0$, i.e.

$$i \frac{\partial g^{(\pm)}(x,t)}{\partial t} = L_0 g^{(\pm)}(x,t), \tag{3.15}$$

where

$$g^{(\pm)}(x,0) = g_0^{(\pm)}(x) = \frac{2}{\pi} \int_0^\infty F^{(\pm)}(k) \frac{\sin kx}{k} k^2 dk . \tag{3.16}$$

If, in analogy with Sec. 2, we introduce a unitary transformation T_0 of $\mathcal{L}_x$ into $\mathcal{L}_k$ given by

$$T_o g = G: \quad G(k) = \int_0^{\infty} g(x) \frac{\sin kx}{k} dx, \tag{3.17}$$

then $g_o^{(\pm)}(x)$ will be expressed in terms of $f_o(x)$ by the formula:

$$g_o^{(\pm)}(x) = T_o^* T^{(\pm)} f_o(x). \tag{3.18}$$

From lemma 3.1 follows

Theorem 3.1. _If $f_o(x)$ is orthogonal to the eigenfunctions of the discrete spectrum of L, then the solution of the Schroedinger equation (3.1) as $t \to \pm \infty$ behaves like the solution of the Schroedinger equation (3.15) with initial data $g_o^{(\pm)}(x)$ given by (3.18), in the sense that_

$$\int_0^{\infty} |f(x,t) - g^{(\pm)}(x,t)|^2 dx \to 0 \text{ as } t \to \pm \infty . \tag{3.19}$$

4. Transformation operators.

In the following we will need the representation of a solution of (1.1) with a potential $q(x)$ in terms of solutions of the equation with other potentials and, in particular, with $q(x) \equiv 0$, i.e. in terms of trigonometric functions.

The simplest expression of this kind can be deduced in the following manner. On the basis of (1.9) for $x \neq 0$, the function $h(x,s) = f(x,s) - e^{is\lambda}$ is square-integrable in s along any line parallel to the real axis, and in the upper half-plane $\int_{-\infty}^{\infty} |h(x, \sigma + i\tau)|^2 d\sigma = O(e^{-2\tau x})$. By a theorem of Titchmarsh,

$$A(x,y) = \frac{1}{2\pi} \int_{-\infty}^{\infty} \left(f(x,k)-e^{ikx}\right) e^{-iky} dk = 0, \quad x > y. \tag{4.1}$$

Inverting this Fourier transform, one thus obtains for f(x,s)

$$f(x,s) = e^{isx} + \int_{x}^{\infty} A(x,y)e^{isy}dy, \quad \tau \geqq 0, \tag{4.2}$$

where A(x,y) is square-integrable in y for $x \neq 0$. With certain modifications, this procedure yields an expression for $\varphi(x,s)$ valid for all s

$$\varphi(x,s) = \frac{\sin sx}{s} + \int_{0}^{x} K(x,y) \frac{\sin sy}{s} dy. \tag{4.3}$$

However, the derivation gives almost no information concerning the kernels K(x,y) and A(x,y). Equivalent equations for these kernels can be deduced by substituting expressions (4.2) and (4.3) for f(x,s) and $\varphi(x,s)$ into (1.6) and (1.5), respectively, and by eliminating the trigonometric functions. Doing this, one obtains the equations

$$K(x,y) = \frac{1}{2} \int_{\frac{x-y}{2}}^{\frac{x+y}{2}} q(t)dt + \int_{\frac{x-y}{2}}^{\frac{x+y}{2}} dt \int_{0}^{\frac{x-y}{2}} dz\; q(t+z)K(t+z,t-z), \quad x \geqq y, \tag{4.4}$$

$$A(x,y) = \frac{1}{2} \int_{\frac{x+y}{2}}^{\infty} q(t)dt - \int_{\frac{x+y}{2}}^{\infty} dt \int_{0}^{\frac{y-x}{2}} dz\; q(t-z)A(t-z,t+z), \quad y \geqq x. \tag{4.5}$$

By then solving these equations by the method of successive approximations, we get the inequalities

$$|K(x,y)| \le \frac{1}{2} \int_{\frac{x-y}{2}}^{x} |q(t)|dt \exp \int_{0}^{\frac{x+y}{2}} t|q(t)|dt \quad , \tag{4.6}$$

$$|A(x,y)| \le \frac{1}{2} \int_{\frac{x+y}{2}}^{\infty} |q(t)|dt \exp \int_{x}^{\infty} t|q(t)|dt \tag{4.7}$$

for $K(x,y)$ and $A(x,y)$. From these integral equations, it also follows that $K(x,y)$ and $A(x,y)$ are differentiable and estimates for their derivatives can be determined. For example,

$$\left|\frac{\partial}{\partial x} A(x,y) + \frac{1}{4} q\left(\frac{x+y}{2}\right)\right| \le K \int_{x}^{\infty} |q(t)|dt \int_{\frac{x+y}{2}}^{\infty} |q(t)|dt \ . \tag{4.8}$$

A similar inequality holds for $\partial A(x,y)/\partial y$.

By virtue of (4.6) and (4.7), the theory of Volterra-type integral equations may be applied to integral equations having $K(x,y)$ and $A(x,y)$ as kernels. Thus, if $g(x)$ belongs to a given class of functions (which we do not specify), then the equations

$$U_B f(x) \equiv f(x) + \int_{0}^{x} K(x,y)f(y)dy = g(x), \tag{4.9}$$

$$V_B f(x) \equiv f(x) + \int_{x}^{\infty} A(x,y)f(y)dy = g(x) \tag{4.10}$$

have solutions which can be represented in the form:

$$f(x) = g(x) + \int_0^x \tilde{K}(x,y)g(y)dy \equiv U_B^{-1}g(x), \tag{4.11}$$

$$f(x) = g(x) + \int_x^\infty \tilde{A}(x,y)g(y)dy \equiv V_B^{-1}g(x). \tag{4.12}$$

In connection with this, the kernels $\tilde{K}(x,y)$ and $\tilde{A}(x,y)$ will have estimates similar to (4.6) and (4.7). A more precise definition of the operators U_B and V_B as operators in Hilbert space will be given in the subsequent sections.

Relationships analagous to (4.3) exist between any two solutions of (1.1) with different potentials. Let $\varphi_1(x,k)$ and $\varphi_2(x,k)$ be solutions of (1.1) with potentials $q_1(x)$ and $q_2(x)$, respectively: two relations of type (4.3) can then be derived:

$$\varphi_1(x,k) = U_B^{(1)} \frac{\sin kx}{k}, \quad \varphi_2(x,k) = U_B^{(2)} \frac{\sin kx}{k}. \tag{4.13}$$

Inverting the second of these and substituting the result in the first, we get

$$\varphi_1(x,k) = U_B^{(1)}(U_B^{(2)})^{-1}\varphi_2(x,k) \tag{4.14}$$

or, explicitly

$$\varphi_1(x,k) = \varphi_2(x,k) + \int_0^x K(x,y)\varphi_2(y,k)dy . \tag{4.15}$$

It is also possible to derive an inequality for $K(x,y)$ similar to (4.6). Finally, one may also relate two different solutions $f_1(x,k)$ and $f_2(x,k)$ but such an expression will not be required in the following.

5. General theory of transformation operators.

We will consider the operator L introduced in Sec. 2 as one of the functional representations of an abstract operator, which will also be denoted by L. The operator L of Sec. 2 will now be denoted by L^x and the space $\mathcal{L}_x$ in which it operates will be called the coordinate representation or x-representation. Another representation to be considered is the so-called momentum or k-representation. This space is of type $\mathcal{L}_k$, and it is defined by the condition that in it L_o is an operator which multiplies an element by k^2:

$$L_o F(k) = k^2 F(k). \tag{5.1}$$

Both representations are related to one another by the unitary transformation T_o introduced in Sec. 3. In other words, to each element $f(x) \in \mathcal{L}_x$ there corresponds an element F(k) belonging to $\mathcal{L}_k$:

$$F(k) = T_o f(x) = \int_0^\infty f(x) \frac{\sin kx}{k} dx \tag{5.2}$$

and each operator A^x in $\mathcal{L}_x$ is converted into an operator $A^k = T_o A^x T_o^*$. For example, in the k-representation, L is given by

$$L^k F(k) = k^2 F(k) + \frac{2}{\pi} \int_0^\infty V(k,\ell) F(\ell) \ell^2 d\ell, \tag{5.3}$$

where

$$V(k,\ell) = \int_0^\infty \frac{\sin kx}{k} q(x) \frac{\sin \ell x}{\ell} dx. \tag{5.4}$$

The passage from the momentum back to the coordinate representation is effected by using the transformation T_o^*; so, for example, the operator which multiplies by a decreasing function $\Omega(k)$ goes into an integral operator in the x-representation with a kernel

$$\Omega(x,y) = \frac{2}{\pi} \int_0^\infty \frac{\sin kx}{k} \Omega(k) \frac{\sin ky}{k} k^2 dk. \tag{5.5}$$

In Sec. 2, the eigenfunction expansion theorem for L was written down in terms of the operator T and involved various spaces. This is inconvenient. In place of T, we will utilize operators acting in one and the same space and which therefore can be given by one of their representations. These operators are defined in the coordinate and momentum representations in the following way:

$$\left.\begin{aligned} U^k &= T_o T^* \quad \text{in the k-representation} \\ U^x &= T^* T_o \quad \text{in the x-representation} \end{aligned}\right\} \tag{5.6}$$

Such operators will be called transformation operators. In all subsequent discussions, they replace completely the eigenfunctions of the continuous spectrum. Instead of a differential equation for the eigenfunctions $\chi(x,k)$, one has for the operator U, determined by them, the equation

$$LU = UL_o. \tag{5.7}$$

The completeness condition and orthogonality of the eigenfunctions in terms of U are expressed by

$$UWU^* = I_c, \quad U^*UW = I \tag{5.8}$$

(see (2.17)). The operator W which appears in these formulas is an operator which multiplies by the function W(k) in the momentum representation. The convenience of (5.7) and (5.8) is in not having to write indices indicating the spaces in which the operators act - all operators are in one and the same representation. In the many problems related to the operator L, the use of different transformation operators is advantageous. Thus, the operators $U^{(\pm)}$ obtained from the transformations $T^{(\pm)}$ by means of (5.6) turn out to be useful in studying the asymptotic behavior of the solution of the Schroedinger equation

$$i \frac{\partial z(t)}{\partial t} = Lz(t), \quad z(t)\Big|_{t=0} = z_0 \tag{5.9}$$

for large $|t|$. The solution can be written in the following form.

$$z(t) = e^{-iLt} z_0, \tag{5.10}$$

and theorem 3.1 can now be reformulated as

Theorem 5.1. <u>If z_0 is orthogonal to the eigenfunctions of the point spectrum of L, then for the solution of the Schroedinger equation the limiting conditions</u>

$$\lim_{t \to \pm\infty} e^{iL_0 t} z(t) = z_{\pm} \tag{5.11}$$

<u>hold where</u>

$$z_+ = U^{(+)*} z_0, \quad z_- = U^{(-)*} z_0. \tag{5.12}$$

If we denote the 'normalizing factor' relating $U^{(+)}$ and $U^{(-)}$ by S:

$$U^{(-)} = U^{(+)} S, \tag{5.13}$$

then z_+ and z_- in formula (5.12) are related to one another by:

$$z_+ = S z_- . \tag{5.14}$$

It is not difficult to show from the definition of $U^{(+)}$ and (5.13) that in the momentum representation this is an operator which multiplies by the function

$$S(k) = \frac{M(-k)}{M(k)} \; . \tag{5.15}$$

$S(k)$ obviously has an absolute value equal to one. From theorem 5.1, it follows that there exists a solution of the Schroedinger equation which is asymptotic to the vector $z_-(t) = e^{-iL_o t} z_-$ as $t \to -\infty$, z_- being an arbitrary element. In addition, this solution for $t = 0$ will be orthogonal to the eigenfunctions of the point spectrum of L and consequently will behave like $e^{-iL_o t} z_+$ as $t \to -\infty$, z_+ being constructed from (5.14) using the operator S. Thus, the general formulation of the scattering problem described in the introduction turns out to be valid for L, and S plays the role of the scattering operator. We have also shown that S is unitary and commutes with the energy operator.

The fundamental aim of the survey is to establish the relationship between L and S. The operator $U_B = I + K$, defined in the coordinate representation by (4.9), will play an important role in this. By the definition of this section, this operator is a transformation operator. In fact, if we assume the transformation operator to be an integral operator, then its kernel in the x-representation is obtained by expanding the corresponding solution $\chi(x,k)$ with respect to $(\sin kx)/k$:

$$U(x,y) = \frac{2}{\pi} \int_0^\infty \overline{\chi(x,k)} \, \frac{\sin ky}{k} \, k^2 dk =$$

$$= \frac{2}{\pi} \int_0^\infty \left(\chi(x,k) - \frac{\sin kx}{k}\right) \frac{\sin ky}{k} \, k^2 dk + \delta(x-y). \tag{5.16}$$

In exactly this way, the kernel of the operator $U_B - I$ with a corresponding solution $\varphi(x,k)$ was deduced in Sec. 4. One can give an exact meaning to this crude argument. In consequence of the fact that the function $M(k)$ is the normalizing factor for the solution $\varphi(x,k)$, with the help of which U_B is constructed, the operator W appearing in the formula for U_B of type (5.8) is given by the function

$$W(k) = \frac{1}{M(k)M(-k)} . \tag{5.17}$$

A characteristic feature of U_B is that it is a Volterra operator in the coordinate representation. This property is closely related to the fact that the potential $q(x)$ is diagonal in this representation (i.e. it is a multiplication operator). In fact, the triangularity of the kernel $K(x,y)$ is a consequence of the fact that $\varphi(x,s)$ is an entire function of s which is due to the diagonal property of the potential. Conversely, let there exist for some operator $L = L_o + V$ a transformation operator U_B of Volterra-type in the x-representation,

$$U_B(x,y) = \delta(x-y) + \eta(x-y)K(x,y) \tag{5.18}$$

where

$$\eta(t) = \begin{cases} 1 & \text{for } t > 0 \\ 0 & \text{for } t < 0. \end{cases} \tag{5.19}$$

From (5.7), we find

$$V(I + K) = KL_o - L_oK. \tag{5.20}$$

The operator $B = KL_o - L_oK$ has the kernel

$$B(x,y) = 2\delta(x-y) \frac{d}{dx} K(x,x) + \eta(x-y) \left[\frac{\partial^2 K(x,y)}{\partial x^2} - \frac{\partial^2 K(x,y)}{\partial y^2} \right] . \tag{5.21}$$

Multiplying (5.20) by $(I + K)^{-1}$, we deduce by virtue of (5.21) that

$$V(x,y) = \delta(x-y)2 \frac{d}{dx} K(x,x) + \eta(x-y)C(x,y) \tag{5.22}$$

and since V has to be self-adjoint, that

$$C(x,y) = 0, \quad V(x,y) = \delta(x-y)2 \frac{d}{dx} K(x,x), \tag{5.23}$$

i.e., L is defined by an equation of type (1.1). A precise statement of the above conclusions is contained in the following theorem:

Theorem 5.2. <u>Let L_o be an operator defined in the x-representation by the differential expression $L_o y = -y''$ and the condition $y(0) = 0$. If there exists a self-adjoint transformation $L = L_o + V$ and Volterra transformation $U_B = I + K$, whose kernel is differentiable and if (5.7) is satisfied, then in the x-representation, V is an operator which multiplies by the function</u>

$$q(x) = 2 \frac{d}{dx} K(x,x). \tag{5.24}$$

If the kernel of the transformation is not differentiable, then it is still possible to define L by the differential expression $Ly = -y'' + q(x)y$; in this case, $q(x)$ is a generalized function such as the derivative of an ordinary function, and in particular, may have δ-function singularities.

From theorem 5.2, we have simultaneously shown that the kernel $K(x,y)$ satisfies the equation

$$q(x)K(x,y) = \frac{\partial^2 K(x,y)}{\partial x^2} - \frac{\partial^2 K(x,y)}{\partial y^2}, \quad x > y. \tag{5.25}$$

In fact, by substituting (5.23) into the left side and (5.21) into the right side of (5.20), we find

$$q(x)\delta(x-y) + \eta(x-y)q(x)K(x,y) = q(x)\delta(x-y)$$

$$+ \eta(x-y)\left[\frac{\partial^2 K(x,y)}{\partial x^2} - \frac{\partial^2 K(x,y)}{\partial y^2}\right], \tag{5.26}$$

and from this follows equation (5.25). Of course, in the general case, when $K(x,y)$ does not have derivatives, it is a generalized solution of this equation. On the basis of this equation for $K(x,y)$ and the conditions (5.24) and $K(x,0) = 0$ (this follows directly from the definition of $K(x,y)$), it is not difficult to derive (4.4), already used earlier.

6. The functions W(k) and S(k), their properties and relationship.

In this section, we will study the functions $W(k)$ and $S(k)$ which were introduced in the preceding section by the formulas:

$$W(k) = \frac{1}{M(k)M(-k)} \quad (-\infty < k < \infty), \tag{6.1}$$

$$S(k) = \frac{M(-k)}{M(k)} \quad (-\infty < k < \infty), \tag{6.2}$$

and we will establish the relationship between them. In addition to the properties of $M(k)$ assembled in lemma 1.6, we still require one more property:

Lemma 6.1. M(k) has the representation

$$M(k) = 1 + \int_0^\infty F(t)e^{ikt}dt, \tag{6.3}$$

where $|F(t)|$ is an integrable function.

The proof follows immediately from the definition of $M(k)$ and the expression (4.2). These imply

$$M(k) = f(0,k) = 1 + \int_0^\infty A(0,t)e^{ikt}dt, \tag{6.4}$$

but $|A(0,t)|$ is integrable because of the inequality (4.7).

From (6.1) and (6.2) and by using Wiener's theorem on the Fourier transform of absolutely integrable functions, one can deduce the following lemmas:

Lemma 6.2. W(k) possesses the following properties:

1) W(k) is a positive even function

$$W(k) > 0, \quad W(-k) = W(k) \quad (-\infty < k < \infty); \tag{6.5}$$

2) W(k) - 1 has an absolutely integrable Fourier transform

$$W(k) = 1 + \int_{-\infty}^{\infty} H(t)e^{ikt}dt = 1 + 2\int_{0}^{\infty} H(t)\cos kt\, dt. \tag{6.6}$$

Lemma 6.3 The function S(k) possesses the following properties:

1) $|S(k)| = S(0) = S(\infty) = 1, \; S(-k) = S(k) = \overline{S(k)}^{-1}$, (6.7)

2) S(k) - 1 has an absolutely integrable Fourier transform:

$$S(k) = 1 + \int_{-\infty}^{\infty} F(t)e^{-ikt}\, dt; \tag{6.8}$$

3) $\arg S(k)\Big|_{-\infty}^{\infty} = -4i\pi m, \quad m \geq 0,$ (6.9)

where m is the number of discrete eigenvalues.

The last property is obtained on the basis of Cauchy's theorem for the number of zeros of an analytic function by taking into account that $\arg S(k) = -2 \arg M(k)$. The properties of the functions W(k) and S(k) enumerated in these lemmas are characteristic. For if they are fulfilled, there exists a unique function M(s) which is analytic and bounded in the upper half-plane, and behaves asymptotically like $M(s) = 1 + o(1)$ for large $|s|$. Furthermore, M(s) has a finite number of simple zeros $s_n = i\kappa_n$ $(n = 1,\dots,m)$, and on the real axis, (6.1) and (6.2) are satisfied. In other words, there exists a function possessing the properties of the function $M(s)$ associated with some operator of type L having m discrete eigenvalues.

Let us first show the uniqueness of the function $M(s)$. Suppose there exist two functions $M_1(s)$ and $M_2(s)$, analytic and bounded in the upper half-plane $\tau > 0$, having there a finite number of zeros at $s_n = i\kappa_n$ and such that

$$S(k) = \frac{M_1(-k)}{M_1(k)} = \frac{M_2(-k)}{M_2(k)} , \tag{6.10}$$

on the real axis, $S(k)$ is a given function satisfying the conditions of lemma 6.3. Then the function $M_1(s)/M_2(s)$ is analytic and bounded in the upper half-plane and real on the real axis so that it has a bounded analytic continuation into the lower halfplane. By Liouville's theorem, it follows that $M_1(s)/M_2(s) = C$. Thus, the asymptotic behavior of M_1 and M_2 as $|s| \to \infty$ implies that $C = 1$, i.e. $M_1 = M_2$. By an analogous argument, one may prove the uniqueness of a function $M(s)$ satisfying (6.1) for given $W(k)$. It is only necessary to consider $\ln(M_1/M_2)$.

We now complete the solution of the stated problem. We begin by reconstructing $M(k)$ from $S(k)$ and consider first the case when $m = 0$ in (6.9). If we normalize the phase $\eta(k) = (i/2) \ln S(k)$ so that $\eta(0) = 0$, then $\eta(\infty) = 0$ and

$$\eta(k) = - \int_0^\infty \gamma(t) \sin kt \, dt, \quad \int_0^\infty |\gamma(t)| dt < \infty \tag{6.11}$$

by the Wiener-Levi theorem. The function

$$M(k) = \exp \int_0^\infty \gamma(t) e^{ikt} \, dt \tag{6.12}$$

is the solution of our problem. Suppose now that $m \neq 0$, the function

$$\tilde{S}(k) = S(k) \prod_{n=1}^{m} \left(\frac{k - i\kappa_n}{k + i\kappa_n} \right)^2 \qquad (6.13)$$

possesses all the properties of S(k) but for it $\tilde{m} = 0$. Therefore the relation

$$S(k) = \frac{\tilde{M}(-k)}{\tilde{M}(k)} \qquad (6.14)$$

holds where $\tilde{M}(s)$ has no zeros in the upper halfplane. The solution of our problem will then be the function

$$M(k) = \tilde{M}(k) \prod_{n=1}^{m} \frac{k - i\kappa_n}{k + i\kappa_n} \quad . \qquad (6.15)$$

Consider now the construction of M(k) for a given W(k). By the Wiener-Levi theorem, the function $\rho(k) = \ln W(k)$ can be represented by

$$\rho(k) = -2 \int_0^\infty \gamma(t) \cos kt \, dt. \qquad (6.16)$$

The function

$$\tilde{M}(s) = \exp \int_0^\infty \gamma(t) e^{ist} dt \qquad (6.17)$$

is analytic in the upper half-plane, has no zeros there, and possesses the right asymptotic behavior for large $|s|$. The solution of our problem will be the function

$$M(k) = \tilde{M}(k) \prod_{n=1}^{m} \frac{k - i\kappa_n}{k + i\kappa_n} \, . \qquad (6.18)$$

In the following, the functions M(k), W(k), and S(k), related to some operator L, will be called the M-, W-, and S-functions of this operator. Any of these functions characterize the spectrum of the corresponding operator L, with M(k) giving the most absolute characterization.

The following assertions are a consequence of the above cited derivation.

1) If L has no discrete spectrum, its S-function is uniquely determined by its W-function according to (6.16), (6.17), and (6.14).

2) If two operators L_1 and L_2 have the same W-function, L_1 has no discrete spectrum, but L_2 has discrete eigenvalues at $\lambda_n = -\kappa_n^2$ $(n = 1, \ldots, m)$, then their M and S-functions are related by

$$M_1(k) = M_2(k) \prod_{n=1}^{m} \frac{k + i\kappa_n}{k - i\kappa_n}, \quad S_1(k) = S_2(k) \prod_{n=1}^{m} \left(\frac{k - i\kappa_n}{k + i\kappa_n}\right)^2 . \tag{6.19}$$

These statements will be used in Secs. 8, 11, and 12.

7. The transformation operator A(x,y).

Together with U_B, in Sec. 4 there was introduced an operator V_B with a kernel in the x-representation given by (4.10). In contrast to U_B, the operator V_B is not a transformation operator in the sense of Sec. 5. In fact, its kernel can not be obtained as the Fourier sine transform of a solution of (1.1) which satisfies a zero boundary condition. The present section is devoted to a clarification of the relationship between U_B and V_B. We will suppose that there is no discrete spectrum. This allows one to use the concise operator notation which was introduced in the preceding sections.

Consider first the operator $\widetilde{U}_B = U_B W$ which is the transformation operator related to the solution

$$\tilde{\varphi}(x,k) = \varphi(x,k)W(k) = \frac{1}{2ik}\left[\frac{f(x,k)}{M(k)} - \frac{f(x,-k)}{M(-k)}\right]. \tag{7.1}$$

The kernel of the operator $\tilde{K} = \tilde{U}_B - I$ in the x-representation is obtained by taking the Fourier transform of $\tilde{\varphi}(x,k) - k^{-1} \sin kx$:

$$\tilde{K}(x,y) = \frac{2}{\pi} \int_0^{\infty} \left(\tilde{\varphi}(x,k) - \frac{\sin kx}{k}\right) \frac{\sin ky}{k} k^2 dk =$$

$$= \frac{1}{2\pi} \int_{-\infty}^{\infty} \left(\frac{f(x,k)}{M(k)} - e^{ikx}\right) e^{-iky} dk. \tag{7.2}$$

By virtue of the analyticity and boundedness of $g(x,s) = f(x,k)e^{-isx}/M(s)$ in the upper halfplane, $\tilde{K}(x,y) = 0$ for $x > y$. Therefore, in the x-representation the operator $\tilde{U}_B$ is given by

$$\tilde{U}_B f(x) = f(x) + \int_x^{\infty} \tilde{K}(x,y) f(y) dy. \tag{7.3}$$

The completeness and orthogonality relations

$$U_B W U_B^* = I \tag{7.4}$$

in terms of $\tilde{U}_B$ are then given by

$$\tilde{U}_B \tilde{W} \tilde{U}_B^* = I, \quad \tilde{W} = W^{-1}, \tag{7.5}$$

and from this,

$$(\tilde{U}_B^*)^{-1} W (\tilde{U}_B)^{-1} = I. \tag{7.6}$$

We note that the operator $(\tilde{U}_B^*)^{-1}$ is a Volterra operator similar to U_B, i.e. the integration goes from 0 to x. In Sec. 9, it will be shown, that a Volterra-type operator is uniquely determined by (7.6). On the basis of this, one can conclude, by comparing (7.4) and (7.6) that

$$(\tilde{U}_B^*)^{-1} = U_B. \tag{7.7}$$

This establishes the relationship between the two operators $\tilde{U}_B$ and U_B. On the other hand, by taking the inverse Fourier transform of

$$h(x,k) = \frac{f(x,k)}{M(k)} = e^{ikx} + \int_x^\infty \tilde{K}(x,y)e^{iky}dy, \tag{7.8}$$

one easily gets the relation between $\tilde{U}_B$ and V_B. Let $\Pi(t)$ denote the function

$$\Pi(t) = \frac{1}{2\pi}\int_{-\infty}^{\infty}\left(\frac{1}{M(k)} - 1\right)e^{-ikt}dk. \tag{7.9}$$

Because $1/M(s)$ is analytic in the upper halfplane,

$$\Pi(t) = 0, \quad t < 0. \tag{7.10}$$

Recalling that

$$f(x,k) = e^{ikx} + \int_x^\infty A(x,y)e^{iky}dy, \tag{7.11}$$

we obtain from (7.8) on the basis of the convolution theorem

$$A(x,y) + \Pi(y-x) + \int_x^y A(x,t)\Pi(y-t)dt = \tilde{K}(x,y). \tag{7.12}$$

This is the sought relation. It will be convenient to write this equation as an operator equation. With this in mind, we associate with $\Pi(t)$ the operator Q which is given in the coordinate representation by:

$$Qf(x) = (I + \Pi)f(x) \equiv f(x) + \int_x^{\infty} \Pi(y-x)f(y)dy. \tag{7.13}$$

By virtue of the boundedness of $1/M(s)$, it is not difficult to show that $Q=I+\Pi$ is a bounded operator in our Hilbert space. The well-known properties of integral equations with difference-type Volterra kernels imply that Q has an inverse

$$P = I + \Gamma = Q^{-1} \tag{7.14}$$

which in the x-representation has a structure similar to Q:

$$Pf(x) = (I + \Gamma)f(x) \equiv f(x) + \int_x^{\infty} \Gamma(y - x)f(y)dy. \tag{7.15}$$

The function $\Gamma(t)$

$$\Gamma(t) = \frac{1}{2\pi} \int_{-\infty}^{\infty} (M(k) - 1)e^{-ikt}dk \tag{7.16}$$

was introduced in the previous section. In terms of the operator Q equation (7.12) takes the form

$$\tilde{U}_B = V_B Q. \tag{7.17}$$

The completeness and orthogonality of the eigenfunctions can now be written down using the operator V_B. Substituting (7.17) in (7.5), we find

$$V_B Q W^{-1} Q^* V_B^* = I. \tag{7.18}$$

Let us clarify the structure of the operator $QW^{-1}Q^*$. For this, the x-representation will be convenient. In this representation, the operator $W^{-1} = \tilde{W}$ has the form $\tilde{W} = I + \tilde{\Omega}$ where $\tilde{\Omega}$ is an integral operator with the kernel

$$\widetilde{\Omega}(x,y) = \frac{2}{\pi} \int_0^\infty \frac{\sin kx}{k} \left(\frac{1}{W(k)} - 1\right) \frac{\sin ky}{k} k^2 dk = \widetilde{H}(x-y) - \widetilde{H}(x+y) \qquad (7.19)$$

and

$$\widetilde{H}(t) = \frac{1}{2\pi} \int_{-\infty}^{\infty} \left(\frac{1}{W(k)} - 1\right) e^{-ikt} dk. \qquad (7\ 20)$$

Denote by H_1 the operator with kernel $\widetilde{H}(x-y)$ and by H_2 the operator with kernel $-\widetilde{H}(x+y)$. Then, in our terms, the identity

$$\frac{1}{W(k)} \frac{1}{M(k)} = M(-k) \qquad (7.21)$$

can be written in the form:

$$(I + \Pi)(I + H_1) = I + \Gamma^*. \qquad (7.22)$$

From this,

$$QW^{-1}Q^* = (I + \Pi)(I + H_1 + H_2)(I + \Pi^*) =$$

$$= (I + \Gamma^* + (I + \Pi)H_2)(I + \Pi^*) = I + (I + \Pi)H_2(I + \Pi^*). \qquad (7.23)$$

One can easily verify that the second term in (7.23) is an integral operator whose kernel is a sum which can be constructed using the function

$$\widetilde{F}(t) = \frac{1}{2\pi} \int_{-\infty}^{\infty} \left(S(k) - \frac{1}{M^2(k)}\right) e^{ikt} dk. \qquad (7.24)$$

For $t > 0$, $\widetilde{F}(t)$ coincides with the function $F(t)$ introduced in the preceding section. We will associate with the latter the operator

$$Ff(x) = \int_0^\infty F(x + y) f(y) dy. \qquad (7.25)$$

We have thus shown that completeness and orthogonality of the eigenfunctions can be written in terms of V_B in the following way:

$$V_B(I - F)V_B^* = I. \tag{7.26}$$

To conclude this section, we note the significance of the various operators used in Hilbert space. By virtue of the uniform boundedness of $M(k)$, $1/M(k)$, $W(k)$, $1/W(k)$, and $S(k)$, the operators M, W, S, W^{-1}, M^{-1}, Q, P, and $I-F$ are bounded in our Hilbert space. All the remaining operators used, to wit U_B, $\tilde{U}_B$, and V_B, can be obtained by multiplying the unitary operator $U^{(+)}$ or $U^{(+)*}$ by one of the above-mentioned operators. Consequently, all of these operators are also bounded operators in our Hilbert space.

8. Integral equations for the kernels $K(x,y)$ and $A(x,y)$.

In this section, we shall establish how the functions $W(k)$ and $S(k)$, as well as the kernels $K(x,y)$ and $A(x,y)$, are connected. The completeness and orthogonality relations for the eigenfunctions written in terms of U_B and V_B are such expressions. However, if we write out these equations in the x-representation, for example, the kernels $K(x,y)$ or $A(x,y)$ will enter non-linearly. This turns out to be inconvenient in solving the inverse problem. Nevertheless, the fact that the operators U_B and V_B are Volterra operators in the x-representation allows one to easily obtain expressions relating $W(k)$ and $S(k)$ and the kernels $K(x,y)$ and $A(x,y)$ in which the latter enter linearly.

For simplicity, it is first assumed that there is no discrete spectrum. Rewrite the equality

$$U_B W U_B^* = I \tag{8.1}$$

in the form

$$U_B W = \tilde{U}_B, \quad \tilde{U}_B = (U_B^*)^{-1}. \tag{8.2}$$

By expressing (8.2) in the x-representation, one obtains

$$K(x,y) + \Omega(x,y) + \int_0^x K(x,t)\Omega(t,y)dt = \tilde{K}(x,y). \tag{8.3}$$

Here

$$\Omega(x,y) = \frac{2}{\pi} \int_0^\infty \frac{\sin kx}{k} (W(k) - 1) \frac{\sin ky}{k} k^2 dk. \tag{8.4}$$

Since $K(x,y) = 0$ for $x > y$, it finally follows that

$$K(x,y) + \Omega(x,y) + \int_0^x K(x,t)\Omega(t,y)dt = 0, \quad x > y. \tag{8.5}$$

In an analogous way, the relation

$$V_B(I - F)V_B^* = I \tag{8.6}$$

gives

$$A(x,y) = F(x+y) + \int_x^\infty A(x,t)F(t+y)dt, \quad x < y, \tag{8.7}$$

$$F(t) = \frac{1}{2\pi} \int_{-\infty}^\infty (S(k) - 1)e^{ikt}dk. \tag{8.8}$$

In case there is a discrete spectrum, similar equations can be gotten by starting directly from the representation (4.3) for $\varphi(x,k)$ and the completeness and orthogonality relations in the form (2.6). Equations for the transformation operator which relates the solutions $\varphi(x,k)$ of two different equations of type (1.1) will now be derived. The initial equations are the relations:

$$\varphi_2(x,k) = \varphi_1(x,k) + \int_0^x K(x,t)\varphi_1(t,k)dt, \tag{8.9}$$

$$\varphi_1(y,k) = \varphi_2(y,k) + \int_0^x \tilde{K}(t,y)\varphi_2(t,k)dt, \tag{8.10}$$

$$\sum_{n_i=1}^{m_i} C_{n_i}\varphi_{n_i}(x)\varphi_{n_i}(y) + \frac{2}{\pi}\int_0^\infty \varphi_i(x,k)W_i(k)\varphi_i(y,k)k^2dk = \delta(x-y) \quad (i = 1,2). \tag{8.11}$$

Multiplication of (8.10) by $\varphi_2(x,k)W_2(k)$ and integration with respect to k gives

$$\frac{2}{\pi}\int_0^\infty \varphi_2(x,k)W_2(k)\varphi_1(y,k)k^2dk + \sum_{n_2=1}^{m_2} C_{n_2}\varphi_1(x,i\kappa_{n_2})\varphi_{n_2}(y) =$$

$$= \delta(x-y) + \int_0^x \tilde{K}(t,y)\delta(x-t)dt = \delta(x-y), \quad x > y. \tag{8.12}$$

Analogous operations with (8.9) now yield

$$\delta(x-y) = \frac{2}{\pi}\int_0^\infty \varphi_1(x,k)W_2(k)\varphi_1(y,k)k^2dk + \sum_{n_2=1}^{m_2} C_{n_2}\varphi_1(x,i\kappa_{n_2})\varphi_1(y,i\kappa_{n_2}) +$$

$$+ \int_0^x K(x,t)\Bigg[\frac{2}{\pi}\int_0^\infty \varphi_1(t,k)W_2(k)\varphi_1(y,k)k^2dk +$$

$$+ \sum_{n_2=1}^{m_2} C_{n_2}\varphi_1(t,i\kappa_{n_2})\varphi_1(y,i\kappa_{n_2})\Bigg]dt, \quad x > y. \tag{8.13}$$

Subtracting (8.11) for $i = 1$ from (8.13), we obtain

$$K(x,y) + \Omega(x,y) + \int_0^{\infty} K(x,t)\Omega(t,y)dt = 0, \quad x > y, \tag{8.14}$$

where

$$\Omega(x,y) = \frac{2}{\pi} \int_0^{\infty} \varphi_1(x,k)\Big[W_2(k) - W_1(k)\Big]\varphi_1(y,k)k^2 dk +$$

$$+ \sum_{n_2=1}^{m_2} c_{n_2}\varphi_1(x,i\kappa_{n_2})\varphi_1(y,i\kappa_{n_2}) - \sum_{n_1=1}^{m_1} c_{n_1}\varphi_{n_1}(x)\varphi_{n_1}(y). \tag{8.15}$$

Since the generalization of (8.7) will not be required later on its derivation is omitted. Equations (8.5) and (8.14) are called the Gelfand-Levitan equations, and (8.7) is called the Marchenko equation.

These equations enable one to solve inverse problems, i.e. reconstruct the operator L in different formulations. We first apply these equations in solving the following two problems.

1) Given the function $W(k)$ possessing the properties enumerated in lemma 6.2. To construct an operator L not having a discrete spectrum for which $W(k)$ is its W-function.

2) Given an operator L_1 with no discrete spectrum, its W-function and the function $q_1(x) \not\equiv 0$. To construct an operator L whose W-function coincides with that of L_1 and which has discrete eigenvalues at m given points $\lambda_n = -\kappa_n^2$ $(n = 1,\dots,m)$.

The basic problem, i.e. reconstruct L from its S-function and its discrete energy levels, will be solvable if these two problems can be solved. For, let there be given a function $S(k)$ possessing the properties enumerated in lemma 6.3 and m distinct positive numbers κ_n $(n = 1,\ldots,m)$. By the procedure described in Sec. 6, a unique $W(k)$ can be constructed from this data. If in conjunction with problem 1, we now construct the operator $\tilde{L}$ without a discrete spectrum for which this $W(k)$ is its W-function then by the assertions of Sec. 6, the S-function of this operator will be

$$\tilde{S}(k) = S(k) \prod_{n=1}^{m} \left(\frac{k - i\kappa_n}{k + i\kappa_n}\right)^2. \tag{8.16}$$

If in accordance with problem 2, we now start with the operator $\tilde{L}_1$ and construct the operator L which has the same W-function as $\tilde{L}_1$ and m discrete eigenvalues at the points $\lambda_n = -\kappa_n^2$, then L will have as its S-function the initially-given function $S(k)$ and the basic problem will, therefore, be solved.

§ 9. Existence of a solution of the Gelfand-Levitan equation. Solution of the first problem.

Let us proceed to solve the problems formulated in the previous section. To do this, we will use the Gelfand-Levitan equations (8.5), (8.4) and (8.14), (8.15) and we therefore first show the existence of solutions.

We go directly to the general equation (8.14). Let the operator L_1 be given. That is, its eigenfunctions $\varphi_1(x,k)$, its W-function, $W_1(k)$, its discrete eigenvalues $\lambda_{n_1} = -\kappa_{n_1}^2$ and the corresponding normalizing factors C_{n_1} are given. Also, let the function $W(k)$ possessing the properties enumerated in lemma 6.2 as well as the arbitrary positive numbers κ_n and C_n be given. Assume that the κ_n are distinct. On the basis of this data, we construct the function $\Omega(x,y)$ by means of (8.15). By virtue of the given conditions, this function will be absolutely integrable with respect to x or y in any finite interval. Equation (8.14) for $K(x,y)$ is an equation with respect to the argument y, the kernel and free term depending on x as a parameter. For fixed x, this equation is a Fredholm equation. Therefore, to show that a unique solution exists, it suffices to prove that the homogeneous equation has only a trivial solution.

Suppose that for fixed x_o, the equation

$$h_o(y) + \int_0^{x_o} h_o(t)\Omega(t,y)dt = 0 \tag{9.1}$$

has a solution. Denote

$$h(y) = \begin{cases} h_o(y) & \text{for } y \leq x_o \\ 0 & \text{for } y > x_o. \end{cases} \tag{9.2}$$

Multiplying (9.1) by $h(y)$ and integrating with respect to y, we obtain

$$\int_0^\infty h^2(y)dy + \int_0^\infty\int_0^\infty h(t)\Omega(t,y)h(y)dtdy = 0, \tag{9.3}$$

or, if we substitute the expression for $\Omega(x,y)$ from (8.15)

$$\int_0^\infty h^2(y)dy + \frac{2}{\pi}\int_0^\infty \left[\int_0^\infty h(y)\varphi_1(y,k)dy\right]^2 \left[W_2(k) - W_1(k)\right]k^2dk +$$

$$+ \sum_{n=1}^{m} C_n \left[\int_0^\infty h(y)\varphi_1(y,i\kappa_n)dy\right]^2 - \sum_{n_1=1}^{m_1} C_{n_1}\left[\int_0^\infty h(y)\varphi_{n_1}(y)dy\right]^2 = 0. \tag{9.4}$$

By the application of Parseval's equality for the system of functions $\varphi_1(x,k)$ this may be written in the form

$$\frac{2}{\pi}\int_0^\infty \left[\int_0^\infty h(y)\varphi_1(y,k)dy\right]^2 W_2(k)k^2dk + \sum_{n=1}^{m} c_n\left[\int_0^\infty h(y)\varphi_1(y,i\kappa_n)dy\right]^2 = 0. \tag{9.5}$$

Since $W_2(k)$ is positive, this implies that $h(y)$ is orthogonal to the subspace corresponding to the continuous spectrum of L. Consequently, $W_2(k)$ is a linear combination of eigenfunctions of the discrete spectrum. But this is impossible since $h(y)$ vanishes identically for $y > x_0$, so that $h(y) \equiv 0$.

The remainder of this section and the two succeeding sections will be devoted to solving the first problem. In this, we will be dealing with the equation

$$K(x,y) + \Omega(x,y) + \int_0^x K(x,t)\Omega(t,y)dt = 0, \quad x > y, \tag{9.6}$$

$$\Omega(x,y) = \frac{2}{\pi}\int_0^\infty \frac{\sin kx}{k}(W(k) - 1)\frac{\sin ky}{k}k^2dk, \tag{9.7}$$

which was obtained from the condition

$$U_B W U_B^* = I \tag{9.8}$$

for the operator $U_B = I + K$. Conversely, we shall show that the operator $U_B = I + K$ constructed with the help of the solution of this equation, possesses the property (9.8), where the function $W(k)$ is only required to satisfy the conditions enumerated in lemma 6.2.

We note that the solution of (9.6) determines a kernel $\tilde{K}(x,y)$ which is different from zero only for $x < y$:

$$\tilde{K}(x,y) = \begin{cases} K(x,y) + \Omega(x,y) + \int_0^x K(x,t)\Omega(t,y)dt, & x < y, \\ 0, & x > y. \end{cases} \tag{9.9}$$

With this kernel, we will associate the operator $\tilde{K}$. Let $\tilde{U}_B = I + \tilde{K}$. In terms of U_B and $\tilde{U}_B$, relation (9.9) becomes

$$U_B W = \tilde{U}_B, \tag{9.10}$$

and to prove (9.8), it remains to be shown that

$$U_B = (\tilde{U}_B^*)^{-1}. \tag{9.11}$$

Since W(k) is real, W is selfadjoint. From (9.10), it follows that $\tilde{U}_B U_B^* = U_B W U_B^*$ so that $\tilde{U}_B U_B^*$ is also selfadjoint. On the other hand, since $\tilde{U}_B$ and U_B^* are Volterra operators, $\tilde{U}_B U_B^*$ is also of Volterra-type and this property together with the selfadjointness implies that $\tilde{U}_B U_B^* = I$. This is what we had to show.

These arguments are of a rather formal character since no estimates for the kernels K(x,y) and $\tilde{K}(x,y)$ were given which would have explained the meaning of U_B and $\tilde{U}_B$ as operators in Hilbert space. However, relation (9.8) when written out in the x-representation only involves integrals over finite intervals of integration (whose convergence it is not necessary to prove). And this relation can be proven starting from (9.6) directly. Anyway, inasmuch as this relation is proven, it is possible to attach meaning to the operator U_B. In fact, the function M(k) constructed from W(k) by the procedure of Sec. 6, is a bounded function and a bounded operator can be associated with it, namely, an operator which multiplies by M(k) in the momentum representation. Since $M(k) \neq 0$, the operator M^{-1} is also bounded. From (9.8), we find that $U^{(+)} = U_B M^{-1}$ is unitary and from this follows the boundedness of U_B.

With the operator thus obtained, we define an operator L by

$$L = U_B L_o U_B^{-1}, \tag{9.12}$$

where L_o is the operator which multiplies by k^2 in the k-representation which was introduced in Sec. 3 and Sec. 5. Since $L = U_B L_o W U_B^*$ by virtue of (9.8), L is selfadjoint because W and L_o commute. On the basis of theorem 5.2, we can assert that L is a differential operator in the x-representation related to an equation of type (1.1) with a potential given by

$$q(x) = 2 \frac{d}{dx} K(x,x). \tag{9.13}$$

Equation (9.8) shows that the given function W(k) is the W-function for L. And so, the first problem of Sec. 8 is solved in the sense that an operator L has been shown to exist. The properties of the potential q(x) occurring in the solution of this problem will be investigated in the following section.

In conclusion, we note that our considerations imply that (9.8) determines a unique Volterra operator U_B. This fact was previously used in Sec. 7.

10. Marchenko's equation. The properties of the potential.

In studying the properties of q(x), it is convenient to use Marchenko's equation

$$A(x,y) = F(x+y) + \int_x^\infty A(x,t)F(t+y)dt, \quad x < y, \tag{10.1}$$

$$F(t) = \frac{1}{2\pi} \int_{-\infty}^{\infty} (S(k) - 1)e^{ikt} dk. \tag{10.2}$$

The existence of a solution of (10.1) will not be proven. Instead, it will be shown that it is equivalent to the Gelfand-Levitan equation (8.5) on the condition, of course, that the corresponding functions W(k) and S(k) are related by the formulas of Sec. 6. In the preceding section, the equivalence of the Gelfand-Levitan equation (8.5) and the relation $U_B W U_B^* = 1$ for the operator $U_B = I + K$ was shown. By repeating the arguments of Sec. 7 and Sec. 8, one will find that $V_B = \tilde{U}_B Q^{-1}$ is an operator which can be constructed using the solution A(x,y) of (10.1) Therefore (10.1) has a solution for $x \geq 0$. Conversely, it is easily shown that if the operator V_B is formed with respect to a solution of (10.1), then $U_B = (Q^* V_B^*)^{-1}$ is the operator formed by using the solution of the Gelfand-Levitan equation (8.5). This implies the uniqueness of a solution of (10.1)

The potential $q(x) = 2 \frac{d}{dx} K(x,x)$ can be simply expressed in terms of A(x,y). Thus, from (7.12) it follws that $A(x,x) = \tilde{K}(x,x) - \Pi(0)$ and $dA(x,x)/dx = dK(x,x)/dx$. If (7.7) is written out in terms of K(x,y) and $\tilde{K}(x,y)$, one obtains

$$\tilde{K}(y,x) = K(x,y) + \int_y^x \tilde{K}(t,x)K(t,y)dt = 0, \tag{10.3}$$

from which it follows that $dK(x,x)/dx = -d\tilde{K}(x,x)/dx$. Finally,

$$q(x) = 2 \frac{d}{dx} K(x,x) = -2 \frac{d}{dx} A(x,x). \tag{10.4}$$

The solution of (10.1) will now be investigated. It turns out that q(x) behaves in many respects like the derivative of F(t) for $t > 0$. We obtain some estimates and first consider how $F'(t)$ depends on $q(x)$. It is convenient to introduce the functions

$$\sigma(x) = \int_x^\infty |q(t)|dt, \quad \sigma_1(x) = \int_x^\infty t|q(t)|dt. \tag{10.5}$$

The inequalities cited in Section 4 for $A(x,y)$ and its partial can now be written·

$$|A(x,y)| \leq K\sigma\left(\frac{x+y}{2}\right) \tag{10.6}$$

$$\left| \frac{\partial}{\partial x} A(x,y) + \frac{1}{4} q\left(\frac{x+y}{2}\right) \right| \leq K\sigma\left(\frac{x+y}{2}\right)\sigma(x). \tag{10.7}$$

In (10.1), set $x = y$. Then

$$A(x,x) = F(2x) + \int_x^\infty A(x,t)F(t+x)dt \tag{10.8}$$

or

$$F(2x) = A(x,x) - 2\int_x^\infty A(x,2t-x)F(2t)dt. \tag{10.9}$$

The last equation is solved by the method of successive approximations, and an estimate for its solution can be gotten

$$|F(2x)| \leq K\sigma(x). \tag{10.10}$$

Differentiation of (10.9) with respect to x and the inequalities (10.7) and (10.10) then imply

$$\left|F'(2x) + \frac{1}{4} q(x)\right| \leq K\sigma^2(x). \tag{10.11}$$

We now study the converse question for which it is convenient to transcribe (10.1) to the form.

$$A(x,x+y) = F(2x+y) + \int_0^\infty A(x,x+t)F(2x+t+y)dt. \qquad (10.12)$$

Let F_x denote the operator

$$F_x g(y) = \int_0^\infty g(t)F(2x+t+y)dt \qquad (10.13)$$

Because (10.1) has a unique solution for all x and since F_x is small for large x, the norm of $(I - F_x)^{-1}$ in $\mathcal{L}_1(0,\infty)$ is uniformly bounded. This implies:

$$\int_0^\infty |A(x,x+y)|dy = \int_x^\infty |A(x,y)|dy \leq K \int_{2x}^\infty |F(t)|dt. \qquad (10.14)$$

Introducing now the notation

$$\tau(x) = \int_x^\infty |F'(t)|dt, \quad \tau_1(x) = \int_x^\infty t|F'(t)|dt, \qquad (10.15)$$

one may write

$$|F(x)| \leq \int_x^\infty |F'(t)|dt = \tau(x). \qquad (10.16)$$

From (10.1) and the inequality (10.14), the uniform estimate

$$|A(x,y)| \leq K\tau(x+y) \qquad (10.17)$$

is gotten. Let us consider the differentiability of A(x,x). Introduce the notation:

$$B(x,y) = A(x,x+y). \tag{10.18}$$

The difference quotient

$$\frac{\Delta_x B(x,y)}{h} = \frac{B(x+h,y) - B(x,y)}{h} \tag{10.19}$$

satisfies an equation of the same kind as $B(x,y)$:

$$\frac{\Delta_x B(x,y)}{h} = \frac{\Delta_x F(2x+y)}{h} + \int_0^\infty B(x,t) \frac{\Delta_x F(2x+t+y)}{h} dt +$$

$$+ \int_0^\infty \frac{\Delta_x B(x,t)}{h} F(2x+2h+t+y)dt. \tag{10.20}$$

The free term

$$\frac{\Delta_x F(2x+y)}{h} + \int_0^\infty B(x,y) \frac{\Delta_x F(2x+t+y)}{h} dt \tag{10.21}$$

can be estimated, uniformly in h, because of the differentiability of $F(t)$, whence follows the differentiability of $B(x,y)$. Moreover, in exactly the same way as before, one obtains the estimate for $\frac{\partial B}{\partial x}(x,y)$:

$$\int_0^\infty |B'_x(x,y)|dy \leq K\tau(2x). \tag{10.22}$$

This implies that

$$\left| \frac{d}{dx} A(x,x) - 2F'(2x) \right| \leq K\tau^2(2x) \tag{10.23}$$

or

$$\left| F'(2x) + \frac{1}{4} q(x) \right| \leq K\tau^2(2x). \tag{10.24}$$

Thus the inequalities (10.11) and (10.24) explicitly show the analogous behavior of $q(x)$ and $F'(2x)$. In particular, they imply that if

$$\int_0^\infty x|q(x)|dx < \infty \tag{10.25}$$

holds, then so does

$$\int_0^\infty x|F'(2x)|dx < \infty \tag{10.26},$$

and conversely. These inequalities also can be used to establish necessary and sufficient conditions on the S-function guaranteeing that the corresponding potential decay like x^{-n}, or exponentially, or vanish identically for $x > A$, etc. A precise formulation of these will not be given here because of the lack of space.

11. Krein's equation. Asymptotic behavior of $\varphi(x,k)$

The eigenfunctions $\varphi(x,k)$ of the operator L are formed using the formula

$$\varphi(x,k) = \frac{\sin kx}{k} + \int_0^x K(x,y) \frac{\sin ky}{k} dy. \tag{11.1}$$

By (1.21), these functions behave asymptotically like

$$\varphi(x,k) \approx \frac{A(k)}{k} \sin(kx - \eta(k)), \tag{11.2}$$

for large x. $A(k)$ and $\eta(k)$ are respectively the modulus and argument of the M-function of L which is uniquely determined by $W(k)$ by means of the relation

$$W(k) = \frac{1}{M(k)M(-k)} \tag{11.3}$$

and the condition of analyticity (see Sec. 6).

It is not easy to get the asymptotic expression (11.2) directly from (11.1). Another representation of $\varphi(x,k)$ will therefore be obtained from which (11.2) can be deduced without difficulty. Of course, this representation will not only serve to prove (11.2), but the procedure outlined below is independently interesting in that it reveals more completely the structure of the kernel $A(x,y)$ of the transformation operator.

The starting point is the Gelfand-Levitan equation

$$K(x,y) + \Omega(x,y) + \int_0^x K(x,t)\Omega(t,y)dt = 0, \quad y < x, \tag{11.4}$$

$$\Omega(x,y) = \frac{2}{\pi} \int_0^\infty \sin kx\left[W(k) - 1\right]\sin ky\, dk. \tag{11.5}$$

From (11.5), it is evident that

$$\Omega(x,y) = H(x-y) - H(x+y), \tag{11.6}$$

where

$$H(t) = \frac{1}{\pi} \int_0^\infty \left[W(k) - 1\right]\cos kt\, dk \tag{11.7}$$

coincides with the function $H(t)$ introduced in Sec. 6 (see (6.6)). If a solution $K(x,y)$ is sought in the form

$$K(x,y) = \Gamma_{2x}(x-y) - \Gamma_{2x}(x+y), \tag{11.8}$$

then one obtains for $\Gamma_{2x}(t)$ the equation

$$\Gamma_{2x}(t) + H(t) + \int_0^{2x} \Gamma_{2x}(s)H(s-t)ds = 0, \tag{11.9}$$

which is equivalent to (11.4). This is an equation for $\Gamma_{2x}(t)$ in the argument t, x playing the role of a parameter. This fact underscores the unsymmetric dependence of $\Gamma_{2x}(t)$ on its arguments. Equation (11.9) is called Krein's equation. Knowing its solution, one can easily find the kernel K(x,y), and together with it, the potential q(x) and the solution $\varphi(x,k)$. However, these functions can be expressed directly in terms of $\Gamma_{2x}(t)$, as follows.

If the resolvent of the kernel H(x-t) on the interval (0,a) is denoted by $G_a(x,t)$:

$$G_a(x,t) + H(x-t) + \int_0^a G_a(x,s)H(s-t)ds = 0, \tag{11.10}$$

then

$$\Gamma_{2x}(t) = G_{2x}(0,t). \tag{11.11}$$

We now note some properties of $G_a(x,t)$. Differentiation of (11.10) with respect to a shows that $G_a(x,t)$ satisfies

$$\frac{\partial}{\partial a} G_a(x,t) = G_a(x,a)G_a(a,t). \tag{11.12}$$

Also,

$$G_a(t,s) = G_a(a-t,a-s). \tag{11.13}$$

In terms of $\Gamma_{2x}(t) = G_{2x}(0,t)$, $q(x) = 2\, dK(x,x)/dx$ is given by

$$q(x) = 2\frac{d}{dx} G_{2x}(0,0) - 2\frac{d}{dx} G_{2x}(0,2x). \tag{11.14}$$

If the notation

$$A(x) = 2G_{2x}(0,2x) = 2\Gamma_{2x}(2x) \tag{11.15}$$

is introduced, then by virtue of (11.12) and (11.13)

$$q(x) = -\frac{d}{dx}A(x) + A^2(x). \tag{11.16}$$

This last formula can be used to reduce the second order differential equation containing $q(x)$ to a system of equations of the first order. Thus,

$$\frac{d^2}{dx^2} - q(x) = \left(\frac{d}{dx} - A(x)\right)\left(\frac{d}{dx} + A(x)\right) \tag{11.17}$$

and the equation

$$-y'' + q(x)y = k^2 y \tag{11.18}$$

is equivalent to the system:

$$\left.\begin{aligned} \frac{d}{dx}y + Ay &= kz, \\ -\frac{d}{dx}z + Az &= ky. \end{aligned}\right\} \tag{11.19}$$

In certain cases, this system turns out to be more convenient than the original equation (11.18). Let us now express the solution $\varphi(x,k)$ in terms of $\Gamma_{2x}(t)$:

$$\varphi(x,k) = \frac{\sin kx}{k} + \int_0^x \Gamma_{2x}(x-y)\frac{\sin ky}{k}dy - \int_0^x \Gamma_{2x}(x+y)\frac{\sin ky}{k}dy =$$

$$= \frac{\sin kx}{k} + \int_0^{2x} \Gamma_{2x}(t)\frac{\sin k(x-t)}{k}dt =$$

$$= \frac{1}{k}\,\mathrm{Im}\left[e^{ikx}\left(1 + \int_0^{2x}\Gamma_{2x}(t)e^{-ikt}dt\right)\right]. \tag{11.20}$$

As $x \to \infty$, we find

$$\varphi(x,k) \to \frac{1}{k} \operatorname{Im} \left[e^{ikx} \left(1 + \int_0^\infty \Gamma(t) e^{-ikt} dt \right) \right], \qquad (11.21)$$

where $\Gamma(t) = \lim_{x \to \infty} \Gamma_{2x}(t)$ is the solution of the equation

$$\Gamma(t) + H(t) + \int_0^\infty \Gamma(s) H(t-s) ds = 0. \qquad (11.22)$$

That is, this equation is satisfied by the function

$$\Gamma(t) = \frac{1}{2\pi} \int_{-\infty}^{\infty} (M(k) - 1) e^{-ikt} dk, \qquad (11.23)$$

introduced in Sec. 6 (see (6.3)). This is not difficult to show if one takes the Fourier transform of the identity (11.3) written out in the form

$$(W(k)-1)(M(k)-1)+(M(k)-1)+(W(k)-1) = \frac{1}{M(-k)} - 1, \qquad (11.24)$$

and takes into account that $1/M(-k)$ is analytic in the lower halfplane. Thus, from (11.21), it follows that

$$\varphi(x,k) \to \frac{1}{k} \operatorname{Im} \left\{ M(-k) e^{-ikx} \right\} \qquad (11.25)$$

as $x \to \infty$ and this, in turn, implies the asymptotic behavior (11.2).

12. Relationship of the operators to the discrete spectra.

In line with our program, we have completed the solution of the first problem and in this section we begin to consider the second problem. The Gelfand-Levitan equation will be considered for the case when the given operator L_1 does not have a discrete spectrum, i.e. all $C_{n_1} = 0$, and the operator being sought has the same function $W(k)$ as L_1 and discrete eigenvalues at $\lambda_n = -\kappa_n^2$ $(n = 1, \ldots, m)$ with corresponding normalizing factors C_n. The equation in this case is given by:

$$K(x,y) + \Omega(x,y) + \int_0^x K(x,t)\Omega(t,y)dt = 0, \quad x > y, \tag{12.1}$$

$$\Omega(x,y) = \sum_{n=1}^{m} C_n\varphi_1(x,i\kappa_n)\varphi_1(y,i\kappa_n). \tag{12.2}$$

This is an equation with a degenerate kernel and is easily solved. For this, it is advantageous to use vector notation. Thus we write $\Omega(x,y)$ in the form:

$$\Omega(x,y) = (\vec{\psi}(x),\vec{\varphi}(y)) \tag{12.3}$$

where $\vec{\psi}(x)$ and $\vec{\varphi}(y)$ are vectors with the components $C_n\varphi_1(x,i\kappa_n)$ and $\varphi_1(y,i\kappa_n)$ respectively. We will seek $K(x,y)$ in the form

$$K(x,y) = (\vec{a}(x),\vec{\varphi}(y)). \tag{12.4}$$

Then the Gelfand-Levitan equation transforms to

$$(\vec{a}(x),\vec{\varphi}(y))+(\vec{\psi}(x),\vec{\varphi}(y))+\left(\int_0^x R(t)dt\vec{a}(x),\vec{\varphi}(y)\right) = 0. \tag{12.5}$$

Here $R(t)$ is the tensor product of the vectors $\vec{\varphi}(t)$ and $\vec{\psi}(t)$, i.e. the matrix with elements

$$r_{ik}(t) = \varphi_i(t)\psi_k(t) \quad (i,k = 1,\ldots,m). \tag{12.6}$$

Let

$$V(x) = I + \int_0^x R(t)dt. \qquad (12.7)$$

Then the general result concerning the existence of a solution of the Gelfand-Levitan equation implies that the matrix V(x) has an inverse for all x. Moreover, from its definition, one can easily verify that this matrix is positive-definite.

By virtue of (12.4) and (12.5), we find

$$K(x,y) = -(V(x)^{-1}\vec{\psi}(x),\vec{\varphi}(y)), \qquad (12.8)$$

and in particular,

$$K(x,x) = -(V(x)^{-1}\vec{\psi}(x),\vec{\varphi}(x)) = -\,sp(V(x)^{-1}R(x)) =$$

$$= -\,sp(V(x)^{-1}\frac{d}{dx}V(x)) = -\frac{d}{dx}\ln\det V(x), \qquad (12.9)$$

so that

$$R(x) = \frac{d}{dx}V(x). \qquad (12.10)$$

On the basis of an investigation of the general Gelfand-Levitan equation (8.14) analogous to that carried out for (8.5), one could show that the operator

$$L = U_B L_1 U_B^{-1}, \qquad (12.11)$$

formed by using the solution K(x,y) of this equation is a differential operator in the x-representation with a potential $q(x) = q_1(x) + \Delta q(x)$ where

$$\Delta q(x) = 2\frac{d}{dx}K(x,x) \quad . \qquad (12.12)$$

Moreover, its eigenfunctions are given by

$$\varphi(x,k) = \varphi_1(x,k) + \int_0^x K(x,y)\varphi_1(y,k)dy, \tag{12.13}$$

its W-function is the function $W(k)$ and the quantities $\lambda_n = -\kappa_n^2$ $(n = 1,\ldots,m)$ are the points of its discrete spectrum. Such an investigation will not be carried out, and the proof of the above assertions for the problem under consideration will be effected by other means which are more simple. The fact that the function $\varphi(x,k)$ given by (12.13) is a solution (1.1) with a potential $q(x) = q_1(x) + \Delta q(x)$ where $\Delta q(x)$ is given by (12.12) will be deduced in Sec. 15. In this section, the properties of these functions will be examined.

From (12.8) and (12.9), one obtains the following formulas for $\Delta q(x)$ and $\varphi(x,k)$:

$$\Delta q(x) = -2\frac{d^2}{dx^2}\ln \| V(x) \|, \quad \| V(x) \| = \det V(x), \tag{12.14}$$

$$\varphi(x,k) = \frac{\left\| \begin{matrix} V(x) & \vec{\psi}(x) \\ \vec{\beta}(x,k) & \varphi_1(x,k) \end{matrix} \right\|}{\| V(x) \|} \tag{12.15}$$

The numerator of this last equation is the determinant of the matrix, composed from the matrix $V(x)$, bordered by the vectors $\vec{\psi}(x)$ and $\vec{\beta}(x,k)$ where

$$\vec{\beta}(x,k) = \int_0^x \vec{\Phi}(y)\varphi_1(y,k)dy. \tag{12.16}$$

The M-function of L can be found from the asymptotic expansion of the solution

$\varphi(x,k)$. We will next investigate this asymptotic behavior and also study the properties of the potential increment $\Delta q(x)$ directly from (12.14) and (12.15) restricting ourselves for simplicity to the case $m = 1$. We will make use of the fact that $\varphi_1(x, i\alpha)$, $\alpha > 0$, behaves asymptotically like

$$\varphi_1(x,i\alpha) = Ne^{\alpha x}[1 + o(1)] \tag{12.17}$$

as $x \to \infty$, and as $x \to 0$,

$$\varphi_1(x,i\alpha) = x[1 + o(1)]. \tag{12.18}$$

Express $\varphi(x,k)$ as the sum of two terms:

$$\varphi(x,k) = \frac{1}{2ik}\,[h(x,k) - h(x,-k)], \tag{12.19}$$

where

$$h(x,k) = M(-k)e^{ikx} + o(1) \tag{12.20}$$

as $x \to \infty$. Equation (12.15) then implies that

$$h(x,k) = h_1(x,k) - \frac{C\varphi_1(x,i\kappa)}{1 + C\int\limits_0^x [\varphi_1(t,i\kappa)]^2 dt} \int\limits_0^x \varphi_1(x,i\kappa)h_1(x,k)dx =$$

$$= M_1(-k)\left[e^{ikx} - \frac{CNe^{\kappa x}}{C\,\frac{N^2}{2\kappa}\,e^{2\kappa x}}\;\frac{N}{\kappa + ik}\,e^{ikx+\kappa x}\right] + o(1) =$$

$$= M_1(-k)\,\frac{k + i\kappa}{k - i\kappa}\,e^{ikx} + o(1), \tag{12.21}$$

so that

$$M(k) = M_1(k)\,\frac{k - i\kappa}{k + i\kappa}\,. \tag{12.22}$$

From this, it follows that L has a discrete eigenvalue at the point $\lambda = -\kappa^2$.

Let us now consider $\Delta q(x)$. By virtue of its definition (12.7), $V(x)$ is twice differentiable even if the potential $q_1(x)$ of the given operator is a generalized function with δ-type singularities. Therefore $\Delta q(x)$ is always locally summable. As $x \to 0$, we have

$$\Delta q(x) = -2\,\frac{d^2}{dx^2}\ln\left\{1 + C\int_0^x t^2[1 + o(1)]\,dt\right\} = -4Cx[1 + o(1)]. \tag{12.23}$$

and as $x \to \infty$

$$\Delta q(x) = -2\,\frac{V''(x)V(x) - (V'(x))^2}{V^2(x)} = -\frac{2}{C}\,(2\kappa)^3\,e^{-2\kappa x}[1 + o(1)]. \tag{12.24}$$

In the general case, $m > 1$, asymptotic forms similar to (12.23) and (12.24) still hold for $\Delta q(x)$ in which $C = \sum_{n=1}^{m} C_n$ in (12.23) and $C = C_r$, $\kappa = \kappa_r$ in (12.24), κ_r being the smallest of the κ_n. In every case,

$$\int_0^\infty x\,|\Delta q(x)|\,dx < \infty\,. \tag{12.25}$$

The solution of the second problem of Sec. 8, except for the one statement to be proven in Sec. 15, is thus completed. The solution is given by (12.14) and (12.15). In these formulas, arbitrary positive numbers may be inserted for the C_n so that we have an m-parameter family of solutions. Formula (12.14) is the formula for the equivalent potentials given by Jost and Kohn.

This simultaneously completes the solution of the basic inverse problem. The results obtained in Secs. 9-12 are summarized in the following theorem.

Theorem 12.1. Every function $S(k)$ possessing the properties:

1) $|S(k)| = S(\infty) = S(0) = 1,$

2) $S(-k) = \overline{S(k)} = S^{-1}(k),$

3) $S(k) = 1 + \int_{-\infty}^{\infty} F(t)e^{-ikt}dt$ with $\int_{-\infty}^{\infty} |F(t)|dt < \infty,$

4) $\arg S(k)\Big|_{-\infty}^{\infty} = -4i\pi m, \quad m \geq 0$

is the S-function of some operator having a continuous spectrum along the half-ray $(0,\infty)$ and m negative eigenvalues. In the x-representation, it is a differential operator of type (1.1) with a potential that may be a generalized function such as the derivative of a locally summable function.

In order that the condition $\int_0^{\infty} x|q(x)|dx < \infty$ be fulfilled, it is necessary and sufficient that $\int_0^{\infty} t|F'(t)|dt < \infty$. If $m > 0$, the potential is not uniquely determined. An m-parameter family of potentials exists such that the associated operators have $S(k)$ as their S-function and the given quantities $\lambda_n = -\kappa_n^2$ as eigenvalues.

13. Operators with M-functions differing by a rational factor.

The general Gelfand-Levitan equation could be solved in closed form for the case treated in the preceding section. This is not the only situation in which this is possible. It turns out that one may explicitly solve the operator equation for the transformation which connects two operators L_1 and L_2 whose M-functions differ by a rational factor. We will consider the case when the given operator L_1 with a potential $q_1(x)$ does not have a discrete spectrum and we will construct the operator L_2 whose M-function is given by

$$M_2(k) = M_1(k) \prod_{\ell=1}^{N} \frac{k + i\alpha_\ell}{k + i\beta_\ell}, \quad \alpha_\ell > 0, \ \beta_\ell > 0. \tag{13.1}$$

The requirement $\beta_\ell > 0$ guarantees that M(k) will be regular in the upper halfplane. The condition $\alpha_\ell > 0$ does not disturb the generality of the discussion since

$$\frac{k - i\alpha}{k + i\beta} = \frac{k - i\alpha}{k + i\alpha}\,\frac{k + i\alpha}{k + i\beta}, \tag{13.2}$$

one may consider the transformation

$$M(k) = M_1(k)\,\frac{k - i\alpha}{k + i\beta}, \tag{13.3}$$

and afterwards carry out the transformation

$$M_2(k) = M_1(k)\,\frac{k - i\alpha}{k + i\alpha}, \tag{13.4}$$

as was done in the preceding section.

We will also assume that all the α_ℓ and β_ℓ are distinct. The equation for the related transformation operator is given by:

$$K(x,y) + \Omega(x,y) + \int_0^x K(x,t)\Omega(t,y)dt = 0, \quad y < x \tag{13.5}$$

$$\Omega(x,y) = \frac{2}{\pi}\int_0^\infty \varphi_1(x,k)\left[W_2(k) - W_1(k)\right]\varphi_1(y,k)k^2dk =$$

$$= \frac{2}{\pi}\int_0^\infty \varphi_1(x,k)W_1(k)\left[\prod_{\ell=1}^{N}\frac{k^2 + \beta_\ell^2}{k^2 + \alpha_\ell^2} - 1\right]\varphi_1(y,k)k^2dk. \tag{13.6}$$

If the term in the brackets is resolved into partial fractions:

$$\prod_{\ell=1}^{N} \frac{k^2 + \beta_\ell^2}{k^2 + \alpha_\ell^2} - 1 = \sum_{\ell=1}^{N} \frac{A_\ell}{k^2 + \alpha_\ell^2}, \tag{13.7}$$

where

$$A_j = \frac{\prod_{\ell=1}^{N} (\beta_\ell^2 - \alpha_j^2)}{\prod_{j\neq\ell}^{N} (\alpha_\ell^2 - \alpha_j^2)}, \tag{13.8}$$

$\Omega(x,y)$ can be expressed in the form

$$\Omega(x,y) = \frac{2}{\pi} \int_0^\infty \varphi_1(x,k) \sum_{\ell=1}^{N} \frac{A_\ell}{k^2 + \alpha_\ell^2} W_1(k)\varphi_1(y,k)k^2 dk = \sum_{\ell=1}^{N} A_\ell g_{\alpha_\ell}(x,y). \tag{13.9}$$

Here, $g_\alpha(x,y)$ is the resolvent kernel of the operator L_1 for $\lambda = -\alpha^2$. In Sec. 2, it was seen that

$$g_\alpha(x,y) = \varphi_1(x,i\alpha) \frac{f_1(y,i\alpha)}{M_1(i\alpha)}, \quad x < y, \quad g_\alpha(x,y) = g_\alpha(y,x), \tag{13.10}$$

is a solution of

$$\left(-\frac{d^2}{dx^2} + \alpha^2 + q_1(x)\right) g_\alpha(x,y) = \delta(x-y). \tag{13.11}$$

Also, any solution of

$$-\psi''(x) - k^2\psi(x) + q_1(x)\psi(x) = 0 \tag{13.12}$$

satisfies

$$(\alpha^2 + k^2) \int_a^b g_\alpha(x,y)\psi(y)dy = \psi(x) + \left[g_\alpha(x,y);\ \psi(y)\right] \Big|_{y=a}^{y=b} , \tag{13.13}$$

where

$$[\varphi(x);\psi(x)] = \varphi'(x)\psi(x) - \varphi(x)\psi'(x). \tag{13.14}$$

We will seek a solution K(x,y) of (13.5) in the form:

$$K(x,y) = \sum_{j=1}^{N} a_j(x)\varphi_j^\beta(y), \tag{13.15}$$

where

$$\varphi_j^\beta(y) = \varphi_1(y,i\beta_j) \quad (j = 1,\ldots,N). \tag{13.16}$$

The substitution of (13.15) into (13.5) gives:

$$\sum_{j=1}^{N} a_j(x)\varphi_j^\beta(y) + \sum_{\ell=1}^{N} A_\ell \varphi_\ell^\alpha(y) \frac{f_\ell^\alpha(x)}{M_\ell^\alpha} +$$

$$+ \sum_{j=1}^{N}\sum_{\ell=1}^{N} A_\ell a_j(x) \int_0^x \varphi_j^\beta(t) g_\ell^\alpha(t,y)dt = 0. \tag{13.17}$$

By (13.13), the last term can be transformed as follows:

$$\sum_{j=1}^{N}\sum_{\ell=1}^{N} A_\ell a_j(x) \int_0^x \varphi_j^\beta(t) g_\ell^\alpha(t,y)dt = \sum_{j=1}^{N}\sum_{\ell=1}^{N} a_j(x) \frac{A_\ell}{\alpha_\ell^2 - \beta_\ell^2} \varphi_j^\beta(y) +$$

$$+ \sum_{j=1}^{N}\sum_{\ell=1}^{N} a_j(x) \left[g_\ell^\alpha(x,y);\ \varphi_j^\beta(y)\right] \frac{A_\ell}{\alpha_\ell^2 - \beta_\ell^2} . \tag{13.18}$$

The first term in this cancels with the first term in (13.17) and the latter may be written:

$$\sum_{\ell=1}^{N} \frac{A_\ell \varphi_\ell^\alpha(y)}{M_\ell^\alpha} \left[f_\ell^\alpha(x) + \sum_{j=1}^{N} W_{\ell j}(x) a_j(x) \right] = 0, \qquad (13.19)$$

where

$$W_{\ell j}(x) = \frac{1}{\alpha_\ell^2 - \beta_j^2} \left[f_\ell^\alpha(x) ; \varphi_j^\beta(x) \right]. \qquad (13.20)$$

It is again advantageous to use vector notation. Due to the linear independence of the φ_ℓ^α, (13.19) is equivalent to the equation

$$\vec{f}(x) + W(x)\vec{a}(x) = 0, \qquad (13.21)$$

from which it follows that

$$\vec{a}(x) = - W^{-1}(x)\vec{f}(x) \qquad (13.22)$$

and

$$K(x,y) = (\vec{a}(x),\vec{\varphi}(y)) = - (W^{-1}(x)\vec{f}(x),\vec{\varphi}(y)). \qquad (13.23)$$

The basic property of the elements of the matrix $W(x)$ is similar to the property (12.10) of the elements of the matrix $V(x)$:

$$\frac{d}{dx} W_{\ell j}(x) = f_\ell^\alpha(x)\varphi_j^\beta(x) . \qquad (13.24)$$

This relation allows one to again obtain an elegant expression for the variation in the potential $\Delta q(x)$ and the solution $\varphi_2(x,k)$:

$$\Delta q(x) = - 2 \frac{d^2}{dx^2} \ln \| W(x) \|, \qquad (13.25)$$

$$\varphi_2(x,k) = \frac{\left\| \begin{matrix} W(x) & \vec{f}(x) \\ \vec{\beta}(x,k) & \varphi_1(x,k) \end{matrix} \right\|}{\| W(x) \|}, \tag{13.26}$$

where

$$\beta_j(x,k) = \frac{\left[\varphi_j^\beta(x);\varphi_1(x,k)\right]}{\beta_j^2 + k^2} = \int_0^x \varphi_j^\beta(t)\varphi_1(t,k)dt. \tag{13.27}$$

The result (13.25) is called Bargmann's formula.

In Sec. 15, it will be shown that $\varphi_2(x,k)$ actually is a solution of (1.1) with the potential $q_2(x) = q_1(x) + \Delta q(x)$. The asymptotic form of $\varphi_2(x,k)$ can be deduced directly from the explicit form of (13.26) just as was done in the preceding section. It is given by

$$\varphi_2(x,k) = \frac{1}{2ik}\left[M_2(-k)e^{ikx} - M_2(k)e^{-ikx}\right] + o(1), \tag{13.28}$$

where

$$M_2(k) = M_1(k) \prod_{\ell=1}^{N} \frac{k + i\alpha_\ell}{k + i\beta_\ell}, \tag{13.29}$$

and this proves the correctness of our solution.

The behavior of $\Delta q(x)$ as $x \to 0$ or $x \to \infty$ can easily be analyzed. For $x \to 0$, $\Delta q(x)$ decreases exponentially as before, however, as $x \to 0$ $\Delta q(x)$ does not tend to zero. Let us consider one factor for the sake of simplicity. On the basis of (13.24)

$$W(x) = W(0) + \int_0^x f_1(x,i\alpha)\varphi_1(x,i\beta)dx. \tag{13.30}$$

But from (13.20), it follows that $W(0) = M_1(i\alpha)/(\alpha^2 - \beta^2)$ and this implies

$$W(x) = \frac{M_1(i\alpha)}{\alpha^2 - \beta^2} \left(1 + (\alpha^2 - \beta^2) \frac{x^2}{2} [1 + o(1)]\right) , \tag{13.31}$$

$$\Delta q(0) = -2(\alpha^2 - \beta^2). \tag{13.32}$$

The formulas cited can be used to determine the potential approximately for given $S(k)$. Any $S(k)$ may be approximated by a rational function $S_R(k)$. The corresponding approximate function $M_R(k)$ will be given by a product

$$M_R(k) = \prod_{\ell=1}^{N} \frac{k + i\alpha_\ell}{k + i\beta_\ell} . \tag{13.33}$$

The potential constructed from this function will be given in terms of trigonometric functions. However, in general, its asymptotic representation differs from that of the exact solution. But in some average interval, the behavior of the potential will be described by the approximate solution rather well.

As an example, take the function

$$M(k) = \frac{k + i\alpha}{k + i\beta} . \tag{13.34}$$

The corresponding phase is given by

$$\cot \eta(k) = \frac{\alpha\beta}{\beta - \alpha} \frac{1}{k} + \frac{1}{\alpha - \beta} k , \tag{13.35}$$

and there is no discrete spectrum. For the potential $q(x)$ one obtains

$$q(x) = 2 \frac{\beta^2(\beta^2 - \alpha^2)}{(\beta \cosh \beta x + \alpha \sinh \beta x)^2} . \tag{13.36}$$

The expression for $\varphi(x,k)$ is more involved and will not be given.

14. The case $\ell > 0$.

Equation (1.1), considered up to this time, is a particular case of

$$L_y^{(\ell)} = - y'' + \left(q(x) + \frac{\ell(\ell + 1)}{x^2}\right) y = s^2 y, \qquad (14.1)$$

which arises when variables are separated in the three-dimensional Schroedinger equation with a spherically-symmetric potential $q(x)$:

$$- \Delta u + qu = s^2 u \qquad (14.2)$$

In this section, all the properties obtained in the preceding sections for the operator L will be carried over to the operator $L^{(\ell)}$. Let

$$\varphi_0^{(\ell)}(x,s) = \frac{1}{s^{\ell+1}} j_\ell(sx), \qquad (14.3)$$

$$f_0^{(\ell)}(x,s) = (i)^{\ell+1} h_\ell^{(1)}(sx), \qquad (14.4)$$

where $j_\ell^{(x)}$, $h_\ell^{(1)}(x)$ are spherical Bessel functions; for any cylindrical function, the corresponding spherical function is given by:

$$z_\ell(x) = \sqrt{\frac{\pi x}{2}}\ Z_{\ell+\frac{1}{2}}(x) \quad (\ell = 0,1,2,\dots). \qquad (14.5)$$

As $x \to 0$ or $x \to \infty$ these functions behave in the following way:

$$\left.\begin{aligned} \varphi_0^{(\ell)}(x,s)\Big|_{x\to 0} &= \frac{x^{\ell+1}}{(2\ell+1)!!}\,[1 + o(1)], \\ \varphi_0^{(\ell)}(x,s)\Big|_{x\to\infty} &= \frac{1}{s^{\ell+1}} \sin\left(sx - \frac{\ell\pi}{2}\right) + o(1), \end{aligned}\right\} \qquad (14.6)$$

$$f_0^{(\ell)}(x,s)\Big|_{x\to 0} = \frac{i^{\ell}(2\ell-1)!!}{(sx)^{\ell}}\left[1 + o(1)\right],$$
$$f_0^{(\ell)}(x,s)\Big|_{x\to\infty} = e^{isx} + o(1). \tag{14.7}$$

They play roles analogous to $\frac{\sin sx}{s}$ and e^{isx} in the case under consideration. The solutions $\varphi^{(\ell)}(x,s)$ and $f^{(\ell)}(x,s)$, which are generalizations of $\varphi(x,s)$ and $f(x,s)$, are determined by the conditions

$$\lim_{x\to 0} \frac{(2\ell+1)!!}{x^{\ell+1}}\,\varphi^{(\ell)}(x,s) = 1, \tag{14.8}$$

$$\lim_{x\to\infty} e^{-isx}\, f^{(\ell)}(x,s) = 1. \tag{14.9}$$

The integral equations, similar to (1.5) and (1.6) and equivalent to (14.1) with the conditions (14.8) and (14.9), are given by:

$$\varphi^{(\ell)}(x,s) = \varphi_0^{(\ell)}(x,s) + \int_0^x J^{(\ell)}(s;x,t)q(t)\varphi^{(\ell)}(t,s)dt, \tag{14.10}$$

$$f^{(\ell)}(x,s) = f_0^{(\ell)}(x,s) - \int_x^\infty J^{(\ell)}(s;x,t)q(t)f^{(\ell)}(t,s)dt. \tag{14.11}$$

Here,

$$J^{(\ell)}(s;x,t) = (is)^{\ell}\left[\varphi_0^{(\ell)}(x,s)f_0^{(\ell)}(t,-s) - \varphi_0^{(\ell)}(t,s)f_0^{(\ell)}(x,-s)\right]. \tag{14.12}$$

By means of these equations, the results of Sec. 1 carry over in respect to the behavior of $\varphi^{(\ell)}(x,s)$ and $f^{(\ell)}(x,s)$ in the complex s-plane, for large x, etc., under the supposition that the potential $q(x)$ satisfies the condition

$$\int_0^\infty x|q(x)|dx < \infty . \qquad (14.13)$$

For example, $\varphi^{(\ell)}(x,s)$ is an entire function of s, $f^{(\ell)}(x,s)$ is analytic in s in the upper halfplane $\tau > 0$, and for these solutions the inequalities

$$|\varphi^{(\ell)}(x,s)| \leq K\left(\frac{x}{1+|s|x}\right)^{\ell+1} e^{|\tau|x}, \qquad (14.14)$$

$$|f^{(\ell)}(x,s)| \leq K\left(\frac{1+|s|x}{|s|x}\right)^{\ell} e^{-\tau x}, \quad t \geq 0, \qquad (14.15)$$

are valid. For large $|s|$, $f^{(\ell)}(x,s)$ behaves asymptotically like

$$f^{(\ell)}(x,s) = e^{isx} + o(1), \quad \tau \geq 0. \qquad (14.16)$$

For real s, $\varphi^{(\ell)}(x,k)$ can be expressed in terms of $f^{(\ell)}(x,k)$:

$$\varphi^{(\ell)}(x,k) = \frac{1}{2ik}\left(\frac{1}{ik}\right)^{\ell}\left[f^{(\ell)}(x,k)M^{(\ell)}(-k)-(-1)^{\ell}f^{(\ell)}(x,-k)M^{(\ell)}(k)\right], \qquad (14.17)$$

where

$$M^{(\ell)}(s) = \lim_{x\to 0} \frac{(sx)^{\ell}}{i^{\ell}(2\ell-1)!!} f^{(\ell)}(x,s) \qquad (14.18)$$

possesses all the properties of the function $M(s)$ enumerated in lemma 1.6, with a correction in the formula for the normalization constant C_n. In the present situation, C_n is given by

$$C_n^{(\ell)} = \int_0^\infty \left[\varphi^{(\ell)}(x,i\kappa_n)\right]^2 dx = \frac{\dot{M}(i\kappa_n)}{2i\kappa_n^{\ell+1}(2\ell+1)!!}\left[\lim_{x\to 0}\frac{f^{(\ell)}(x,i\kappa_n)}{x^{\ell+1}}\right]^{-1}. \quad (14.19)$$

As before, it is assumed that $M^{(\ell)}(0) \neq 0$ since the condition $M^{(\ell)}(0) = 0$ is again equivalent to instability and this restriction is of no consequence. With this assumption, the solution $\varphi^{(\ell)}(x,0)$ behaves asymptotically for large x like

$$\varphi^{(\ell)}(x,0) = Ax^{\ell+1}[1 + o(1)]. \quad (14.20)$$

The results concerning the expansion theorem, the existence of transformation operators, and the asymptotic expansions for large time of the solution of the Schroedinger equation all go over to the present case. Thus, the completeness relation for the eigenfunctions is given by

$$\sum_{n=1}^{m} C_n\varphi_n(x)\varphi_n(y) +$$

$$+ \frac{2}{\pi}\int_0^\infty \varphi^{(\ell)}(x,k)\frac{1}{M^{(\ell)}(k)M^{(\ell)}(-k)}\varphi^{(\ell)}(y,k)k^{2(\ell+1)}dk = \delta(x-y). \quad (14.21)$$

Furthermore, the following representations for $\varphi^{(\ell)}(x,s)$ and $f^{(\ell)}(x,s)$ hold:

$$\varphi^{(\ell)}(x,s) = \varphi_0^{(\ell)}(x,s) + \int_0^x K^{(\ell)}(x,t)\varphi_0^{(\ell)}(t,s)dt, \quad (14.22)$$

$$f^{(\ell)}(x,s) = f_0^{(\ell)}(x,s) + \int_x^\infty A^{(\ell)}(x,t)f_0^{(\ell)}(t,s)dt. \quad (14.23)$$

The scattering operator associated with $L^{(\ell)}$ is constructed with the help of the function

$$S^{(\ell)}(k) = \frac{M^{(\ell)}(-k)}{M^{(\ell)}(k)} . \tag{14.24}$$

This same function occurs in the asymptotic representation of the normalized eigenfunction $\psi^{(\ell)}(x,k) = (ik)^{\ell}\varphi^{(\ell)}(x,k)/M^{(\ell)}(k)$:

$$\psi^{(\ell)}(x,k)\Big|_{x\to\infty} = \frac{1}{2ik}\left[S^{(\ell)}(k)e^{ikx} - (-1)^{\ell}\, e^{-ikx}\right] + o(1). \tag{14.25}$$

As before, these results allow one to solve the inverse problem, i.e. the reconstruction of $L^{(\ell)}$ from its S-function. Thus the kernel $K^{(\ell)}(x,y)$ satisfies the equation

$$K^{(\ell)}(x,y) + \Omega^{(\ell)}(x,y) + \int_0^x K^{(\ell)}(x,t)\Omega^{(\ell)}(t,y)dt = 0, \quad x > y, \tag{14.26}$$

where

$$\Omega^{(\ell)}(x,y) = \sum_{n=1}^{m} c_n\varphi_0^{(\ell)}(x,i\kappa_n)\varphi_0^{(\ell)}(y,i\kappa_n) +$$

$$+ \frac{2}{\pi}\int_0^\infty \varphi_0^{(\ell)}(x,k)\,\frac{1}{M^{(\ell)}(k)M^{(\ell)}(-k)}\,\varphi_0^{(\ell)}(y,k)k^{2(\ell+1)}dk. \tag{14.27}$$

By a repetition of the earlier argument, this equation can be shown to have a unique solution. Also, the completeness relation (14.21) for $\varphi^{(\ell)}(x,k)$, constructed from the solution $K^{(\ell)}(x,y)$ by means of (14.22), can be proven, and a differential equation of type (14.11) can be derived for $\varphi^{(\ell)}(x,k)$. However, tracing the relationships between $S^{(\ell)}(k)$ and the potential $q(x)$ using this equation or the analog of the Marchenko equation turns out to be difficult. This is due to the fact that the kernel of equations of type (14.26) is constructed using Bessel functions for which no simple addition formula exists as for the trigonometric functions. Of course, conditions like (6.7) and (6.9) for the S-function are still valid in the present case. As to the analog of condition (6.8), it is not clear beforehand that formulating it in terms of the Fourier transform of $S^{(\ell)}(k)$ is convenient. One might think that the Fourier transform arose in a natural way in the case $\ell = 0$ because we dealt with trigonometric functions. All the more surprising is the fact that the behavior of the S-function for $L^{(\ell)}$ turns out to be no different than that of the S-function for $L^{(0)}$. We have in particular

Theorem 14.1. *If* $S(k)$ *is the S-function for the operator* $L^{(\ell)}$ *associated with the differential equation (14.1) with a potential* $q(x)$, *then it is also the S-function of an operator* $L^{(m)}$ *with arbitrary* $m = 0,1,2,\ldots,\ell+1,\ldots$ *where the corresponding potential* $q^{(m)}(x)$ *behaves like the potential* $q(x)$ *as* $x \to 0$ *and* $x \to \infty$.

The proof of this theorem will be given in the following section. Also, the sense in which the behavior of the potentials $q(x)$ and $q^{(m)}(x)$ is analogous will be made more precise there.

15. Transformation of Sturm-Liouville type equations.

In the preceding sections, it was repeatedly mentioned that a series of assertions would be proven in Sec. 15 on the basis of a certain general method. The essence of this method will first be presented from which the assertions will then follow.

Let $y_o(x)$ be some particular solution of

$$-y'' + q(x)y = \lambda y \tag{15.1}$$

for $\lambda = \lambda_o$ which does not vanish in the neighborhood of the point $x = a$. Consider the expression

$$y_1(x,\lambda) = \frac{[y(x,\lambda);y_o(x)]}{(\lambda-\lambda_o)y_o(x)}, \tag{15.2}$$

where $y(x,\lambda)$ is an arbitrary solution of (15.1) and $[\varphi;\psi]$ is the Wronskian

$$[\varphi;\psi] = \varphi'(x)\psi(x) - \varphi(x)\psi'(x) . \tag{15.3}$$

Lemma 15.1. <u>The function</u> $y_1(x,\lambda)$ <u>is a solution of (15.1) with a potential</u> $q_1(x) = q(x) + \Delta q(x)$ <u>where</u>

$$\Delta q(x) = -2 \frac{d}{dx} \frac{y_o'(x)}{y_o(x)} = -2 \frac{d^2}{dx^2} \ln y_o(x). \tag{15.4}$$

To prove this, we note that any two arbitrary solutions of (15.1) satisfy the equality:

$$\frac{d}{dx} \frac{[y(x,\lambda_1);y(x,\lambda_2)]}{\lambda_1 - \lambda_2} = y(x,\lambda_1)y(x,\lambda_2). \tag{15.5}$$

Therefore,

$$y_1'(x,\lambda) = y(x,\lambda) - \frac{[y(x,\lambda);y_o(x)]}{(\lambda - \lambda_o)y_o^2(x)} y_o'(x) = y(x,\lambda) - y_1(x,\lambda)v(x), \tag{15.6}$$

where the function $v(x) = y_o'(x)/y_o(x)$ is a solution of the Riccati equation

$$v'(x) + v^2(x) = q(x) - \lambda_o. \tag{15.7}$$

Differentiating (15.6) once and using (15.7), we find

$$y_1''(x,\lambda) = y'(x,\lambda) - y_1'(x,\lambda)v(x) - y_1(x,\lambda)v'(x) =$$

$$= y'(x,\lambda) - y(x,\lambda)\frac{y_o'(x)}{y_o(x)} + y_1(x,\lambda)v^2(x) - y_1(x,\lambda)v'(x) =$$

$$= -\frac{\left[y(x,\lambda), y_o(x)\right]}{y_o(x)} + y_1(x,\lambda)\left[v^2(x) + v'(x)\right] - 2y_1(x,\lambda)v'(x) =$$

$$= \left[\lambda_o - \lambda + v^2(x) + v'(x) - 2v'(x)\right]y_1(x,\lambda),$$

i.e.

$$-y_1''(x,\lambda) + q_1(x)y_1(x,\lambda) = \lambda y_1(x,\lambda) \tag{15.8}$$

This proves the lemma.

Formula (15.2) makes sense for $\lambda \neq \lambda_o$. If $\lambda = \lambda_o$, one of the solutions of the transformed equation is the function

$$z_{10}(x) = \frac{1}{y_o(x)} . \tag{15.9}$$

As a second linearly independent solution, one may take

$$y_{10}(x) = z_{10}(x)\int^x \frac{dt}{z_{10}^2(t)} = \frac{1}{y_o(x)}\int^x y_o^2(t)dt. \tag{15.10}$$

Of course, the solutions (15.9) and (15.10) can be deduced from (15.2) by a limit process. Conversely, the function $y(x,\lambda)$ is expressible in terms of solutions of the transformed equation. For this it suffices to note that (15.9) implies

$$\frac{y_o'(x)}{y_o(x)} = -\frac{z_{10}'(x)}{z_{10}(x)} \tag{15.11}$$

and, therefore, (15.6) can be written in the form:

$$y(x,\lambda) = y_1'(x,\lambda) - y_1(x,\lambda)\,\frac{z_{10}'(x)}{z_{10}(x)} = \frac{[y_1(x,\lambda);z_{10}(x)]}{z_{10}(x)}\,. \qquad (15.12)$$

This expression defines the transformation inverse to (15.2).

Till now, we have considered the transformation (15.2) in the neighborhood of $x = a$ where the mapping solution y_0 does not vanish. We will now describe the behavior of the solution of the transformed equation when this condition does not hold. Having in mind its application to (1.1) and (14.1), we will assume that the singular point occurs at $x = 0$ and that in the neighborhood of $x = 0$, the given potential has a singularity of the type

$$q(x) = \frac{\ell(\ell+1)}{x^2} + O\left(\frac{1}{x^{2-\epsilon}}\right), \quad \epsilon > 0. \qquad (15.13)$$

This requirement is somewhat stronger than the condition imposed on the potential in Sec. 1 and Sec. 14, namely

$$\int_0^\infty x|q(x)|dx < \infty\,, \qquad (15.14)$$

but it **simplifies** considerably all of the calculations. For this situation, (15.1) will be said to have an ℓ-singularity at $x = 0$. Two types of solutions exist in the neighborhood of $x = 0$: one regular

$$y(x,\lambda) = C(\lambda)x^{\ell+1}\left[1 + O(x^\epsilon)\right], \qquad (15.15)$$

and the other irregular

$$z(x,\lambda) = \frac{D(\lambda)}{x^\ell}\left[1 + O(x^\epsilon)\right]. \qquad (15.16)$$

Evidently, all regular solutions only differ by a factor.

Consider now the transformation generated using some regular solution $y(x,\lambda_0)$. The term

$$\Delta q(x) = -2\frac{d^2}{dx^2}\ln x^{\ell+1}[1 + O(x^{\epsilon})] = \frac{2(\ell+1)}{x^2} + O\left(\frac{1}{x^{2-\epsilon}}\right) \qquad (15.17)$$

is added to the potential $q(x)$, so that the equation has an ℓ-singularity at $x = 0$. In deriving (15.17), the asymptotic representation is differentiated but this can be rigorously justified.

Let us observe what happens to a regular and irregular solution under our transformation. By virtue of (15.6) and (15.15)

$$\frac{[y(x,\lambda);y(x,\lambda_0)]}{\lambda - \lambda_0} = \int_0^x y(t,\lambda)y(t,\lambda_0)dt \qquad (15.18)$$

Therefore,

$$y_1(x,\lambda) = C(\lambda)\frac{\int_0^x x^{2(\ell+1)}[1 + O(x^{\epsilon})]dx}{x^{\ell+1}} = \frac{C(\lambda)}{2\ell+3}x^{\ell+2}[1 + O(x^{\epsilon})]. \qquad (15.19)$$

Furthermore,

$$[z(x,\lambda);y(x,\lambda_0)] = (2\ell+1)D(\lambda)C(\lambda_0)[1 + O(x^{\epsilon})] \qquad (15.20)$$

so that

$$z_1(x,\lambda) = \frac{(2\ell+1)D(\lambda)}{x^{\ell+1}}[1 + O(x^{\epsilon})]. \qquad (15.21)$$

In the following, (15.2) will be used to transform a regular solution and (15.12) to transform an irregular one. It has thus been shown that transforming with a regular solution increases the ℓ-singularity at the origin by unity. Moreover, regular and irregular go into regular and irregular solutions respectively. It is easily seen that transforming with an irregular solution lowers the ℓ-singularity of the equation by unity but again preserving the regularity property of the respective solutions.

Thus, to obtain an equation having the same ℓ-singularity as the given equation, two successive transformations must be performed, the first using a regular solution of the given equation and the second then using an irregular solution of the transformed equation. It turns out that the transformations constructed in Sec. 12 and Sec. 13 by means of the Gelfand-Levitan equation could be deduced in this way. We will illustrate this by transforming two operators whose spectra differ by a single eigenvalue. Consider an equation of type (1.1) with a potential satisfying condition (1.2) and assume that none of the points in the discrete spectrum of the associated operator L lie to the left of $\lambda = -\beta_0^2$. Take some $\beta > \beta_0$. Under the given conditions, $\varphi_0(x) = \varphi(x, i\beta)$ vanishes only at $x = 0$. Perform the transformation (15.2) using this solution. Then the transformed equation will have an ℓ-singularity at $x = 0$ with $\ell = 1$. The function

$$\psi_1(x) = \frac{1}{\varphi_0(x)} \left[1 + C \int_0^x \varphi_0^2(t)dt \right] \tag{15.22}$$

will be an irregular solution of the transformed equation. Performing a transformation with this solution, we arrive at an equation which has no singularity at $x = 0$. Let us calculate the potential and solution $\varphi_2(x,k)$ associated with this equation. To do this, it is necessary to combine the formulas

$$\varphi_1(x,k) = \frac{[\varphi(x,k);\varphi_0(x)]}{(k^2 + \beta^2)\varphi_0(x)} = \frac{1}{\varphi_0(x)} \int_0^x \varphi(t,k)\varphi_0(t)dt, \tag{15.23}$$

$$\varphi_2(x,k) = \frac{[\varphi_1(x,k);\psi_1(x)]}{\psi_1(x)}, \tag{15.24}$$

$$\Delta q(x) = -2 \frac{d^2}{dx^2} \ln \varphi_o(x) - 2 \frac{d^2}{dx^2} \ln \psi_1(x). \tag{15.25}$$

This leads to the following results:

$$\varphi_2(x,k) = \varphi(x,k) - \frac{\varphi_o(x)}{1 + C \int_o^x \varphi_o^2(t)dt} \int_o^x \varphi(t,k)\varphi_o(t)dt, \tag{15.26}$$

$$\Delta q(x) = -2 \frac{d^2}{dx^2} \ln \left(1 + C \int_o^x \varphi_o^2(t)dt\right) . \tag{15.27}$$

These formulas agree with (12.14) and (12.15). Thus by a direct verification it has been shown that the function $\varphi_2(x,k)$ determined by these formulas is a solution of (1.1) with the potential given by (15.25).

To justify the unproven assertion of Sec. 13, it is necessary to perform the following: Transform the given equation using the solution $\varphi(x,i\beta)$. Then transform the resulting equation using the solution $f_1(x,i\alpha)$ obtained from the solution $f(x,i\alpha)$ of the given equation by the first transformation. By combining the formulas

$$\varphi_1(x,k) = \frac{[\varphi(x,k);\varphi(x,i\beta)]}{(k^2+\beta^2)\varphi(x,i\beta)} = \frac{1}{\varphi(x,i\beta)} \int_o^x \varphi(t,k)\varphi(t,i\beta)dt, \tag{15.28}$$

$$f_1(x,i\alpha) = \frac{[f(x,i\alpha);\varphi(x,i\beta)]}{(\beta^2-\alpha^2)\varphi(x,i\beta)} = \frac{W(x)}{\varphi(x,i\beta)} , \tag{15.29}$$

$$\varphi_2(x,k) = \frac{[\varphi_1(x,k);f_1(x,i\alpha)]}{f_1(x,i\alpha)} , \tag{15.30}$$

$$\Delta q(x) = -2\frac{d^2}{dx^2}\ln\varphi(x,i\beta) - 2\frac{d^2}{dx^2}\ln f_1(x,i\alpha). \tag{15.31}$$

one finds that the function

$$\varphi_2(x,k) = \varphi(x,k) - \frac{f(x,i\alpha)}{W(x)}\frac{[\varphi(x,k);\varphi(x,i\beta)]}{k^2+\beta^2} \tag{15.32}$$

is the solution of (1.1) with the potential

$$q_2(x) = q(x) - 2\frac{d^2}{dx^2}\ln W(x) = q(x) - 2\frac{d^2}{dx^2}\ln\frac{[f(x,i\alpha);\varphi(x,i\beta)]}{\beta^2-\alpha^2}. \tag{15.33}$$

This proves the statement of Sec. 13.

These properties of a transformation will now be used to prove theorem 14.1. A transformation using any regular solution changes an ℓ-singularity only at $x = 0$. The behavior of the potential increment $\Delta q(x)$ as $x \to \infty$ may be different depending on the location of the parameter λ in the complex plane. For example, in the cases considered till now, the potential increment decreased exponentially as $x \to \infty$. However, there exist solutions which change the singularity of the equation in the same way both for $x \to 0$ and $x \to \infty$. Such is the solution of equation (15.1) with $\lambda = 0$. If it is assumed that

$$q(x) = \frac{\ell(\ell+1)}{x^2} + O\left(\frac{1}{x^{2+\delta}}\right), \quad \delta > 0, \tag{15.34}$$

as $x \to \infty$, then using equations like (14.10) and (14.11), one can show that the regular solution has the asymptotic representation

$$y(x,0) = Cx^{\ell+1}\left[1 + O\left(\frac{1}{x^\delta}\right)\right]. \tag{15.35}$$

as $x \to \infty$ and among the irregular solutions there exists a $z(x,0)$ such that

$$z(x,0) = \frac{D}{x^{\ell}}\left[1 + O\left(\frac{1}{x^{\delta}}\right)\right] \tag{15.36}$$

as $x \to \infty$. Under transformations using these solutions,

$$\Delta q(x) = -2\frac{d^2}{dx^2}\ln y(x,0) = \frac{2(\ell + 1)}{x^2} + O\left(\frac{1}{x^{2+\delta}}\right) , \tag{15.37}$$

$$\Delta q(x) = -2\frac{d^2}{dx^2}\ln z(x,0) = -\frac{2\ell}{x^2} + O\left(\frac{1}{x^{2+\delta}}\right) . \tag{15.38}$$

Thus they have the requisite property of changing the singularity of the equation equally for $x \to 0$ and $x \to \infty$. Let us see how the S-functions change under transformations using these solutions. It is easily verified that the functions

$$f^{(\ell+1)}(x,k) = \frac{\left[f^{(\ell)}(x,k);y(x,0)\right]}{-iky(x,0)} , \tag{15.39}$$

$$f^{(\ell-1)}(x,k) = \frac{\left[f^{(\ell)}(x,k);z(x,0)\right]}{-ikz(x,0)} \tag{15.40}$$

have the asymptotic behavior

$$f^{(\ell\pm1)}(x,k) = e^{ikx} + o(1) \tag{15.41}$$

as $x \to \infty$. Thus they are solutions of the transformed equations of type $f^{(\ell)}(x,k)$. The function $M(k)$, determining the S-function, arises in the asymptotic formula for solutions of type $f^{(\ell)}(x,k)$ as $x \to 0$:

$$f^{(\ell)}(x,k)\Big|_{x\to 0} = \frac{(2\ell - 1)!!}{(kx)^{\ell}}\, i^{\ell} M(k)\left[1 + O(x^{\epsilon})\right]. \tag{15.42}$$

However, by virtue of (15.20), we find

$$f^{(\ell+1)}(x,k)\Big|_{x\to 0} = \frac{(2\ell + 1)!!}{(kx)^{\ell+1}}\, i^{\ell+1} M(k)\left[1 + O(x^{\epsilon})\right], \tag{15.43}$$

$$f^{(\ell-1)}(x,k)\Big|_{x\to 0} = \frac{(2\ell - 3)!!}{(kx)^{\ell-1}}\, i^{\ell-1} M(k)\left[1 + O(x^{\epsilon})\right]. \tag{15.44}$$

Thus, it has been shown that under transformations using the solutions $y(x,0)$ and $z(x,0)$, M and, consequently, the S-function does not change. This assertion together with formulas (15.17), (15.37) and (15.38) for the asymptotic behavior of the potential as $x \to 0$ and $x \to \infty$, proves theorem 14.1.

APPENDIX

(Comments and notes on the literature)

1. The proofs of lemmas 1.1 - 1.3 and 1.5 are found in Levinson's paper [6]. Certain statements are proven in papers [27] and [28]. The case $M(0) = 0$ is detailedly treated by Marchenko and Agranovich [29], [30].

2. The completeness theorem for the eigenfunctions of the operator L in the form (2.6) is proven by Levinson [6] in the absence of a discrete spectrum. The general case is considered by Jost and Kohn [9].

3. Many papers have been devoted to the question of how the solution of the time dependent Schroedinger equation behaves for large $|t|$ both in physics and mathematics. A nonrigorous proof of the existence of limits of the operator $U(0,t) = e^{iLt}e^{-iL_0 t}$ as $t \to \pm\infty$, typical of physics papers, is cited for example, in the survey of Gellmann and Goldberger [31]. From mathematical work, it is necessary to mention first of all the articles of Friedrichs [23], who proved the existence of the limits $U(0,\pm\infty)$. He also showed that the limiting operators are unitary for a wide class of unperturbed operators L_0 on the assumption that the perturbation operator V is small. A formal presentation of his method is cited in the paper of Moses [32]. Cook [33] proved the existence of the limits $U(0,\pm\infty)$ for the three-dimensional operator $-\Delta u + q(x)u$ in all space assuming only that $q(x)$ is square integrable in all space. However, he did not study the question of whether the operator $S = U(0,\infty)^*U(0,-\infty)$ is unitary. The restriction that the perturbation operator V be small is removed in the paper of Ladizhenskaya and Faddeyev [34] using the formalism of Friedrichs.

Theorem 3.1 does not follow from the results of these papers for the restrictions we have imposed on the potential $q(x)$. The elementary proof cited makes use of the concrete properties of the example under consideration and does not carry over to other problems.

4. Povzner [35] and Levitan [36] first obtained and used the representation (4.3) for $\varphi(x,k)$. Formula (4.2) and $f(x,k)$ was deduced by Levin [37] and our first derivation repeats his argument. The method of deriving the integral equations (4.4) and (4.5) as well as the inequalities (4.7) and (4.8) is due to Agranovich and Marchenko [29], [30].

The theorem of Titchmarsh mentioned can be formulated in the following way: A necessary and sufficient condition for $\Phi(x)$ to be the limit of some function $\Phi(z) = \Phi(x + iy)$, which is analytic in the upper halfplane and such that

$$\int_{-\infty}^{\infty} | \Phi(x + iy)|^2 dx = O(e^{-2ky}),$$

is that

$$\varphi(t) = \frac{1}{2\pi} \int_{-\infty}^{\infty} \Phi(x) e^{-ixt} dx = 0, \quad t < k.$$

5. The general concept of transformation operator, as already noted, was developed by Friedrichs [22], [23]. Some of the notation and the proof of theorem 5.2 was taken from the articles of Kay and Moses [18], [20] who applied Friedrichs' method in solving inverse problems.

From the formula (5.15) for the S-function, it follows that $S(k)$, in general, cannot be continued into the complex k-plane. Thus, Heisenberg's supposition that the discrete energy levels in the example in question might be determined by the analytic continuation of the S-matrix is not justified. This fact was noted by Jost [27].

The differential equation (5.25) and the condition (5.24) is the lead-off point for the proof of the existence of the kernel $K(x,y)$ in the paper of Gelfand and Levitan [12]. One easily obtains the integral equation (4.4) from this equation. Chudov [64] proposed using the nonlinear equation gotten from (5.25) by replacing $q(x)$ by $2dK(x,x)/dx$ to solve the inverse problem. Giving the S-function for large x provides Cauchy data for this equation.

6. The method of relating $W(k)$ and $S(k)$ on the basis of the Wiener-Levi theorem was due to Krein [16], [38]. The Wiener-Levi theorem can be stated in the following way. Let the function $\Phi(z)$ be analytic in a region D and then let $F(\lambda)$ be so chosen that the curve $z = F(\lambda)$ $(-\infty \leq \lambda \leq \infty)$ lies inside D. If $F(\lambda)$ is representable in the form

$$F(\lambda) = C + \int_{-\infty}^{\infty} f(t)e^{i\lambda t}dt$$

where $f(t)$ is absolutely integrable, then $\Phi(F(\lambda))$ also possesses this property.

Formula (6.9) was deduced by Levinson [6] and is called Levinson's formula.

7. The cited relationship between the kernels $K(x,y)$ and $A(x,y)$ does not appear in the literature.

8. The first derivation of (8.5) is taken from the paper of Kay and Moses [18]. The derivation of the general equation (8.14) repeats the arguments of Gelfand and Levitan [12].

9. The existence proof for (8.14) is taken from the paper of Jost and Kohn [13] and to a great extent follows the reasoning of Gelfand and Levitan. The subsequent presentation with certain modifications reproduces the arguments of Kay and Moses [18].

10. The analysis of the properties of the potential $q(x)$ is taken from the monograph of Agranovich and Marchenko [30]. The various relationships between $W(k)$ or $S(k)$ and $q(x)$ were obtained by Neuhaus [39]. Friedman [40], Jost [41], and Newton [42]. However the more general results follow from (10.11) and (10.24).

11. Krein's methods are given in a series of articles [43], [44], [16], [45] in DAN SSSR (see also the lectures presented at MGU in 1956-1957). Only certain of his results are cited in the survey. The system of differential equations (11.19) is the starting point of Krein's methods.

12. Formula (12.14) for the increment in the potential was obtained by Jost and Kohn [13]. The simplest formula for the solution $\varphi(x,k)$ of the type (12.15) was due to Krein [16] (for the case $m = 1$).

The portion of theorem 12.1 relating to necessary and sufficient conditions is due to Marchenko and Agranovich [29], [30].

13. In solving (13.5), we have followed the paper of Fulton and Newton [46] who refer to the work of Bargmann as the source of the method used. Expression (13.25) is called Bargmann's formula. Formulas for $\varphi(x,k)$ such as (13.26) are cited by Theiss [47].

Another approach to the problem is developed by Krein for $M_1(k) = 1$. The results obtained are decisively formulated in [48], the formulas being simpler than (13.25) and (13.26). However, their generalizations to the case $M_1(k) \neq 1$ are not deduced.

14. The basic properties of solutions of (14.1) for $\ell > 0$ are obtained by Levinson [6], Jost and Kohn [13], and Newton [49]. The papers of Stashevskaya [50] and Volk [51] are devoted to carrying over the results of Gelfand and Levitan to equations with a singularity at $x = 0$. Theorem 14.1 is due to Marchenko (it was presented at the April, 1956 meeting of the Kharkov mathematical society).

15. A transformation of the type (15.2) was applied for the first time by Crum [52], who used it to change a differential operator over a finite interval into an operator having one less eigenvalue than the original operator. Krein extended Crum's method and applied the results to get a complete characterization of the spectral function of an equation with the singularity $\ell(\ell + 1)/x^2$ in the potential at $x = 0$. Marchenko made use of an analogous transformation to analyze the relationship between the S-function and a potential given by

$$q(x) = q_1(k) + \frac{\ell(\ell + 1)}{x^2}, \quad \int_0^\infty x^{1+\epsilon} |q_1(x)| dx < \infty .$$

The presentation in the survey differs somewhat from the methods of the abovementioned authors.

It is interesting that inasmuch as we verify formulas (12.14), (12.15) and (13.25), (13.26) by algebraical means without reference to the theory of the general Gelfand-Levitan equation, they still hold for complex values of the parameters κ_n, C_n, α_ℓ and β_ℓ. The associated potential is, generally speaking, a complex function with a singularity of the type $m(m + 1)/(x - x_0)^2$ at those points where $||V(x)|| = 0$ or $||W(x)|| = 0$. Here, m is the multiplicity of any such existing zero. This fact was noted by Krein [16] and Theiss [47].

We now briefly consider some of the generalizations of the problem investigated in our survey. By analyzing many of the formulas in the text, one sees that with appropriate changes they remain valid for systems of equations, i.e. for the matrix equation generalizing (1.1):

$$- Y'' + Q(x)Y = k^2 Y .$$

Here, $Q(x)$ is a real symmetric matrix. The solutions $\varphi(x,k)$, $f(x,k)$ and the functions $W(k)$, $M(k)$, and $S(k)$ become matrices in this case. Therefore it is necessary to pay attention to the order of factors in generalizing formulas to the matrix case. The matrix $M(a)$ is analytic in the upper half-plane $\tau > 0$ and singular at those points corresponding to the discrete spectrum. The matrices $W(k)$ and $S(k)$ are related to it by the formulas

$$W(k) = M(k)^{-1} M^{\tau}(-k)^{-1}, \quad S(k) = M(-k) M^{\tau}(k)^{-1},$$

$M^{\tau}(k)$ being the transposed matrix. Similar systems were studied by Jost and Newton [53], Krein [45], and Agranovich and Marchenko [29], [30]. A fundamental difficulty arises in carrying over the discussion of Sec. 6 to the matrix case. In consequence of the noncommutativity of the matrices, the formulas cited there no longer hold. Finding how the matrices $W(k)$ and $S(k)$ are related reverts to studying integral equations of the type:

$$K(t) = F(t) + \int_0^{\infty} F(t + s)K(s)ds.$$

Marchenko and Agranovich deduced necessary and sufficient conditions on the S-matrix so that it corresponds to a matrix potential $Q(x)$ from a given class making use of analogous integral equations. The formulation of conditions directly in terms of the S-matrix still remains an unsolved problem.

Newton [49] and Agranovich and Marchenko [54], [30] considered a system in which the potential has the singularity $\ell_\alpha(\ell_\alpha + 1)\delta_{\alpha\beta}/x^2$; Agranovich and Marchenko in studying such a system reduced it to a regular one by a transformation generalizing the kind described in Sec. 15.

The inverse problem for a system has mainly been treated with the aim of seeing what means are necessary for solving the inverse problem for the Schroedinger equation

$$- \Delta u + q(x)u = k^2 u$$

in all of space when the potential decreases in all directions. However, this problem essentially differs from those treated till now. In fact, the S-matrix in this case is determined by the so-called scattering amplitude $f(k;\alpha,\beta)$ depending on the wave number $k(0 \leq k < \infty)$ and two unit vectors α and β. Thus the S-matrix depends on a larger number of parameters than the potential $q(x)$ which may be regarded as a function of the distance r $(0 \leq r < \infty)$ and one unit vector. In this sense, the problem is overdetermined and it is necessary to look for nontrivial properties of the S-matrix which would decrease the amount of parameters on which it depends.

The simplest problem where an analogous situation occurs is in reconstructing a decreasing potential in the one-dimensional Schroedinger equation:

$$-y'' + q(x)y = k^2 y \quad (- \infty < x < \infty)$$

from its S-matrix. In this case, the S-matrix is a 2-by-2 matrix:

$$S(k) = \begin{pmatrix} s_{11}(k) & s_{12}(k) \\ s_{21}(k) & s_{22}(k) \end{pmatrix} = \begin{pmatrix} a(k) & b(k) \\ b(k) & c(k) \end{pmatrix},$$

and due to the condition of unitariness, is determined by giving three real functions of k $(0 \leq k < \infty)$. The potential can be regarded as being given by two real functions of x $(0 \leq x < \infty)$.

The inverse problem for this case was considered by Kay and Moses [20], [21] (as an example illustrating their general approach to the inverse problem) and by the author [55]. In [55] it is shown, that an additional condition on the S-matrix follows from the analyticity of the coefficient b(k) in the upper halfplane $\tau > 0$. This condition implies that every S-matrix (and potential) is determined by one of the coefficients a(k) or c(k) which may be chosen as an arbitrary function. The reconstruction of the equation with an arbitrary potential over the entire axis from its spectral matrix function was treated by Bloch [63].

The elements of the S-matrix are also analytic in the three-dimensional case. The proof of this fact is given in the papers of Khuri [56] and the author [57] in connection with the so-called dispersion relations.

It is interesting to note in this connection, the statement of the three-dimensional inverse problem as proposed by Moses [58]: The potential q(x) has to be determined from its back scattering amplitude $g(k,\alpha) = f(k;\alpha,-\alpha)$ where α is a vector running over a hemisphere. These two pieces of data, namely two real functions of k $(0 \leq k < \infty)$ and α, which runs over a hemisphere, contain as many parameters as does the potential. It is very probable that the procedure of Moses converges for sufficiently small $g(k,\alpha)$ which in other respects may be a quite arbitrary function.

There exists a series of papers in which the inverse problem is solved for relativistic equations when the latter reduce to ordinary differential equations. The equation obtained by separating variables in the Klein-Gordon equation has been studied by Corinaldesi [59]. The one-dimensional Dirac equation was considered by Kay and Moses [60], Toll and Prats [61] and Verdi [65]. In all of these papers, a relationship is established between the potential and asymptotic phase, the latter being given for both positive and negative energies. This data, just as in the problems described above, depends on a larger number of parameters than does the potential. A correct formulation of the problem for the radial relativistic equation till now has not been given.

Acknowledgement

The translator would like to express his thanks to Dr. Irvin Kay for having read the manuscript, and for his helpful criticisms.

REFERENCES

[1] W. Heisenberg, Die beobachtbaren Grössen' in der Theorie der Elementarteilchen, Zs. f. Phys. 120 (1943), 513-538; 673-702.

[2] C. Møller, General properties of the characteristic matrix in the theory of elementary particles, Kgl. Danske Videnskab Selskab mat. - fys. medd. 23, No. 1 (1945), 22, No. 19 (1946).

[3] C.E. Frødberg, Calculation of the interaction between two particles from the asymptotic phase, Phys. Rev. 72 (1947), 519.

[4] E.A. Hylleraas, Calculation of a perturbing central field of force from the scattering phase shift, Phys. Rev. 74 (1948), 48.

[5] V. Bargmann, Remarks on the determination of a central field of force from the elastic scattering phase shift, Phys. Rev. 75 (1949), 301.

[6] N. Levinson, On the uniqueness of the potential in a Schroedinger equation for a given asymptotic phase, Kgl. Danske Videnskab Selskab mat.-fys. medd. 25, No. 9 (1949).

[7] V.A. Marchenko, Some questions in the theory of one-dimensional linear differential operators of the second order, Part I; Trudi Mosk. matem. o-va 1 (1952), 327-420.

[8] V.A. Marchenko, Some questions in the theory of a differential operator of the second order, Dokl. AN 72 (1950), 457-460.

[9] R. Jost and W. Kohn, Construction of a potential from the phase shift, Phys. Rev. 87 (1952), 979.

[10] R. Jost and W. Kohn, Equivalent potentials, Phys. Rev. 88 (1952), 382.

[11] B. Holmberg, A remark on the uniqueness of the potential determined from the asymptotic phase, Nuovo Cimento 9 (1952), 597.

[12] I.M. Gelfand and B.M. Levitan, On the determination of a differential equation from its spectral function, Izv. AN, ser. matem. 15 (1951), 309-360.

[13] R. Jost and W. Kohn, On the relationship between the phase shift, energy levels and the potential, Kgl Danske Videnskab Selskab mat.-fys. medd. 27, No. 9 (1953).

[14] N. Levinson, A certain explicit relation between the phase shift and scattering potential, Phys. Rev. 89 (1953), 755-757.

[15] M.G. Krein, On the transition function for a one-dimensional boundary value problem of the second order, Dokl. AN 88 (1953), 405-408.

[16] M.G. Krein, On the determination of the potential of a particle from its S-function, Dokl. AN 105 (1955), 433-436.

[17] V.A. Marchenko, Construction of the potential energy from the phase of the scattered waves, Dokl AN 104 (1955), 433-436.

[18] I. Kay and H.E. Moses, The determination of the scattering potential from the spectral measure function, I, Nuovo Cimento 2 (1955), 917.

[19] I. Kay and H.E. Moses, The determination of the scattering potential from the spectral measure function, II, Nuovo Cimento 3 (1956), 67.

[20] I. Kay and H.E. Moses, The determination of the scattering potential from the spectral measure function, III, Nuovo Cimento 3 (1956), 277.

[21] I. Kay and H.E. Moses, The determination of the scattering potential from the spectral measure function, IV, Suppl. al. vol. 5 del Nuovo Cimento (1957), 230-243.

[22] K.O. Friedrichs, Uber die Spectralzerlegung eines Integraloperators, Math. Ann. 115 (1938), 249-272.

[23] K.O. Friedrichs, On the perturbation of continuous spectra, Comm. Pure Appl. Math. 1 (1948), 361-406.

[24] R.G. Newton, Electron scattering by a deuteron, Phys. Rev. 105 (1957), 763-764.

[25] R.G. Newton, Relation of the deuteron photodisintegration cross-section to the neutron-proton force, Phys. Rev. 107 (1957), 1025.

[26] T. Fulton and R.G. Newton, Phenomenological neutron-proton potentials, Phys. Rev. 107 (1957), 1102.

[27] R. Jost, Uber die falschen Nullstellen der Eigenwerte der S-Matrix, Helv. Phys. Acta 20 (1947) 256.

[28] V. Bargmann, On the connection between the phase shift and scattering potential, Rev. Mod. Phys. 21 (1949), 488.

[29] Z.S. Agranovich and V.A. Marchenko, On the construction of the potential from the scattering matrix for a system of differential equations, Dokl. AN, 113 (1957), 951-954.

[30] Z.S. Agranovish and V.A. Marchenko, The Inverse Problem in the Quantum Theory of Scattering, 1959.

[31] M. Gellmann and M. Goldberger, The formal theory of scattering. Phys. Rev. 91 (1953), 398.

[32] H.E. Moses, The scattering operator and the adiabatic theorem, Nuovo Cimento 1 (1955), 103-131.

[33] J.M. Cook, Convergence to the Møller wave-matrix, Journ. of Math. and Phys. 36 (1957), 83.

[34] O.A. Ladizhenskaya and L.D. Faddeyev, On the perturbation of a continuous spectrum, Dokl AN 120 (1958), 1187-1190.

[35] A.Ya. Povzner, On Sturm-Liouville type differential operators along a half-line, Matem. sborn. 23 (1948), 3-52.

[36] B.M. Levitan, Application of generalized displacement operators to linear differential equations of second order, UMN IV (1949), 3-112.

[37] B.Ya. Levin, Fourier and Laplace type transformations using solutions of a differential equation of second order, Dokl AN 106 (1956), 187-190.

[38] M.G. Krein, Integral equations on a half-line with difference type kernels, UMN XIII (1958), 3-110.

[39] M.G. Neuhaus, Determination of the asymptotic behavior of the function q(x) from the properties of the spectral function of the operator $- y'' + q(x)y$, Dokl. AN 102 (1955), 25-28

[40] B. Friedman, On the properties of a singular Sturm-Liouville equation determined by its spectral function, Michigan Math. Journ. 4 (1957), 137-147.

[41] R. Jost, Eine Bermerkung uber den Zusammenhang von Streuphase und Potential, Helv. Phys. Acta 29 (1956), 410-418.

[42] R.G. Newton, Remarks on a scattering theory, Phys. Rev. 101 (1956), 1588-1596.

[43] M.G. Krein, On an effective method for solving the inverse problem, Dokl. AN 94 (1953), 987-990.

[44] M.G. Krein, On integral equations generating differential equations of the second order, Dokl. AN 97 (1954), 21-24.

[45] M.G. Krein, On the theory of accelerants and the S-matrix of canonical differential systems, Dokl. AN 111 (1956), 1167-1170.

[46] T. Fulton and R.G. Newton, Explicit non-central potentials and wave functions for a given S-matrix, Nuovo Cimento 3 (1956), 677-717.

[47] W.R. Theis, Potentiale zu gegebenen Streufunctionen geladener Teilchen mit Absorption, Zs. Naturforsch. 11a (1956), 889-891.

[48] M.G. Krein, On the continuum analog of a formula of Kristoffel in the theory of orthogonal polynomials, Dokl. AN 113 (1957), 970-973.

[49] R.G. Newton, Connection between the S-matrix and the tensor force, Phys. Rev. 100 (1955), 412-428.

[50] V.V. Stashevskaya, On inverse problems of spectral analysis for one class of differential equations, Dokl. AN 93 (1953), 409-411.

[51] V.Ya. Volk, An inversion formula for a differential equation with a singularity at x = 0, UMN VIII, (1953), 141-151.

[52] M.M. Crum, Associated Sturm-Liouville systems, Quart. S. Math. 6 (1955), 121-127.

[53] R.G. Newton and R. Jost, The construction of potentials from the S-matrix for systems of differential equations, Nuovo Cimento 1 (1955), 590-622.

[54] Z.S. Agranovich and V.A. Marchenko, Construction of the tensor force from the given scattering, Dokl. AN 118 (1958), 1055-1058.

[55] L.D. Faddeyev, On the connection between the potential and S-matrix for the one-dimensional Schroedinger operator, Dokl. AN 121 (1958), 63-66.

[56] N.N. Khuri, Analysis of the Schroedinger scattering amplitude and non-relativistic dispersion relations, Phys. Rev. 107 (1957), 1148-1156.

[57] L.D. Faddeyev, On the dispersion relations in the non-relativistic theory of scattering, Zh ETF 35 (1958), 433-439.

[58] H.E. Moses, Calculation of the scattering potential from the reflection coefficient, Phys. Rev. 102 (1956), 559-567.

[59] E. Corinaldesi, Construction of potentials from the phase shift and binding energies of relativistic equations, Nuovo Cimento 11 (1954), 468-478.

[60] H.E. Moses, Calculation of the scattering potential for the one-dimensional Dirac equation from the reflection coefficient and point eigenvalues, Bull. Amer. Phys. Soc. 4 (1957), 240.

[61] J.S. Toll and F. Prats, Connection between the phase shift and potential for the Dirac equation, Phys. Rev. 113, No. 1 (1959.

[62] De Alfaro, On the inversion problem for a Klein-Gordon wave equation, Nuovo Cimento 10 (1958), 675-682.

[63] A.Sh. Bloch, On the determination of a differential equation from its spectral matrix function, Dokl. AN 92 (1953), 209-212.

[64] L.A. Chudov, On a method for constructing a bounded complex potential from its asymptotic phase, Izd. OI Ya I (1958).

[65] M. Verdi, The inversion problem in wave mechanics and dispersion relations, Nuclear Phys. 9 (1958/59), 255-266.

[66] G. Borg, Uniqueness theorems in the spectral theory of $y'' + (\lambda - q(x))y = 0$, Onzieme congres des mathematiciens scandinaves tenu à Trondheim le 22-25 aônt 1949, 276-287.

DISTRIBUTION LIST FOR RESEARCH REPORTS

Contract No. AF 19(604)3495

(ONE copy unless otherwise noted)

Commander
Air Research and Development Command
Washington 25, D.C.
ATTN Major E. Wright, RDTCC

Director
Air University Library
Maxwell Air Force Base, Alabama

Commander
Air Force Missile Test Center
Patrick Air Force Base, Florida
ATTN: Technical Library, MU-411

Tactical Air Group
Directorate of Research and Dev., DCS/D
Headquarters, USAF
Washington, D.C.
ATTN: Major R.I. Stell

Director, Communications and Electronics
HQ U.S. Air Force
Washington 25, D.C.
ATTN: AFOAC S/E

Commander
Air Technical Intelligence Center
Wright-Patterson Air Force Base, Ohio
ATTN: AFCIN-4Bia

Commander
Wright Air Development Center
Wright-Patterson Air Force Base, Ohio
ATTN: Paul Springer, WCLRE-5

Commander
Rome Air Development Center
Griffiss Air Force Base, New York
ATTN: RCSSTL-1

Commander
Air Force Missile Development Center
Holloman Air Force Base, New Mexico
ATTN: HDOIL, Technical Library

Director
Evans Signal Laboratory
Belmar, New Jersey
ATTN: Mr. O.C. Woodyard

U.S. Army Signal Engineering Labs.
Evans Signal Laboratory
Belmar, New Jersey
ATTN: Technical Document Center

Commander
Redstone Arsenal
Huntsville, Alabama
ATTN: Technical Library

Commanding General
U.S. Army Signal Engineering Labs.
Fort Monmouth, New Jersey
ATTN SIGFM/EL-AT

Department of the Army
Office of the Chief Signal Officer
Washington 25, D.C.
ATTN: SIGRD-7

Department of the Army
Office of the Chief Signal Officer
Engineering and Technical Division
Washington 25, D.C.
ATTN: SIGNET-5

(10)Armed Services Technical Information Agency
Document Service Center
Arlington Hall Station
Arlington 12, Virginia
ATTN: TICSCP

(2)Library
Boulder Laboratories
National Bureau of Standards
Boulder, Colorado

National Bureau of Standards
Department of Commerce
Washington 25, D.C.
ATTN Dr. A.G. McNish

(2)Office of Technical Services
Department of Commerce
Washington 25, D.C.
ATTN Technical Reports Section

(2)Hq Air Force Cambridge Research Center
Laurence G. Hanscom Field
Bedford, Massachusetts
ATTN CROTLR-2 - P. Condon

(5)Hq. Air Force Cambridge Research Center
Laurence G. Hanscom Field
Bedford, Massachusetts
ATTN: CROTLS - J. Armstrong

Hq Air Force Cambridge Research Center
Laurence G. Hanscom Field
Bedford, Massachusetts
ATTN: CRRD - R.E. Hiatt

Director, Avionics Division (AV)
Bureau of Aeronautics
Department of the Navy
Washington 25, D.C.

Chief, Bureau of Ships
Department of the Navy
Washington 25, D.C.
ATTN: Mr. E. Johnston, Code 833E

Commander
U.S. Naval Air Missile Test Center
Point Mugu, California
ATTN Code 366

Commander
Air Research and Development Command
ATTN: RDTR
Andrews Air Force Base
Washington 25, D.C.

U.S. Naval Ordnance Laboratory
White Oak
Silver Spring 19, Maryland
ATTN Library

Commander
U.S. Naval Ordnance Test Station
China Lake, California
ATTN: Code 753

Librarian
U.S Naval Postgraduate School
Monterey, California

Director
U.S. Naval Research Laboratory
Washington 25, D.C.
ATTN: Code 2027

Chief of Naval Research
Department of the Navy
Washington 25, D.C.
ATTN: Code 427

Commanding Officer and Director
U.S. Naval Electronics Laboratory (Library)
San Diego 52, California

Department of the Navy
Bureau of Aeronautics
Technical Data Division, Code 4106
Washington 25, D.C.

Bell Telephone Laboratories, Inc.
Whippany Laboratory
Whippany, N.J.
ATTN Technical Information Library

Boeing Airplane Company
Pilotless Aircraft Division
P.O. Box 3707
Seattle 24, Washington

Boeing Airplane Company
Seattle Division
Seattle 14, Washington
ATTN E.T. Allen, Library Supervisor

Convair, A Div. of General Dynamics Corp.
San Diego 12, California
ATTN: Mrs. Dora B. Burke, Engineering Lib

Cornell Aeronautical Laboratory, Inc.
4455 Genesee Street
Buffalo 21, New York
ATTN: Librarian

Douglas Aircraft Company, Inc.
P.O. Box 200
Long Beach 1, California
ATTN: Engineering Library (C-250)

Sylvania Electric Prod. Inc.
Electronics Defense Laboratory
P.O. Box 205
Mountain View, California
ATTN: Library

Emerson and Cuming, Inc.
869 Washington Street
Canton, Massachusetts
ATTN Mr. W.R. Cuming

Federal Telecommunication Labs.
500 Washington Avenue
Nutley 10, N.J.
ATTN Technical Library

General Electric Company
Electronics Park
Syracuse, New York
ATTN: Documents Library, B. Fletcher
Bldg. 3-143A

Hughes Aircraft Co.
Antenna Department
Microwave Laboratory
Building 12, Room 2617
Culver City, California
ATTN: M.D. Adcock

Hughes Aircraft Company
Florence and Teale Streets
Culver City, California
ATTN: Dr. L.C. Van Atta, Assoc. Director
Research Labs.

Hycon Eastern, Inc.
75 Cambridge Parkway
Cambridge, Massachusetts
ATTN: Mrs. Lois Seulowitz
Technical Librarian

Jansky and Bailey, Inc.
1339 Wisconsin Avenue, N.W.
Washington 7, D.C.
ATTN: Mr. Delmer C. Ports

Electromagnetic Research Corp.
711 14th Street, N.W.
Washington 5, D.C.

The Glenn L. Martin Company
Baltimore 3, Maryland
ATTN: Engineering Library
Antenna Design Group

Mathematical Reviews
190 Hope Street
Providence 6, R.I.

Page Communications Engineers, Inc.
710 Fourteenth Street, Northwest
Washington 5, D.C.
ATTN: Librarian

Pickard and Burns, Inc.
240 Highland Avenue
Needham 94, Massachusetts
ATTN: Dr. J.T. DeBettencourt

Milton Keynes UK
Ingram Content Group UK Ltd.
UKHW020824180224
437992UK00005B/213

9 781019 413197